Engineering Education and Practice

NOTRE DAME STUDIES IN ETHICS AND CULTURE

series editor, David Solomon

ENGINEERING

EDUCATION

Embracing a

AND PRACTICE

Catholic

Vision

Edited by

James L. Heft, S.M.

and

Kevin Hallinan

Foreword by David J. O'Brien

University of Notre Dame Press
Notre Dame, Indiana

Manufactured in the United States of America

Library of Congress Cataloging-in-Publication Data

Engineering education and practice : embracing a Catholic vision / edited by
James Heft and Kevin Hallinan ; foreword by David J. O'Brien.
p. cm. — (Notre Dame studies in ethics and culture)
Proceedings of a conference held Sept. 22–24, 2005 at the University of Dayton.
Includes bibliographical references and index.
ISBN-13: 978-0-268-03110-7 (pbk. : alk. paper)
ISBN-10: 0-268-03110-X (pbk. : alk. paper)
1. Engineering—Study and teaching—United States. 2. Engineers—
Education—United States. 3. Engineering—Moral and ethical aspects—United
States. 4. Catholic Church—Education. I. Heft, James. II. Hallinan, K. P.

T73.E447   2011
620.0071'173—dc23

2011036539

∞ *The paper in this book meets the guidelines for permanence and
durability of the Committee on Production Guidelines for
Book Longevity of the Council on Library Resources.*

Contents

PART 3.   INTERNATIONAL SERVICE LEARNING

PART 4.   FORMATION AND PREPARATION OF STUDENTS

Foreword

DAVID J. O'BRIEN

For three days in September 2005, I had the good fortune to participate in a conference on "The Role of Engineering at a Catholic University" at the University of Dayton. I am no engineer, but I am interested
in Catholic higher education, and I found the proceedings extremely
encouraging. My fellow participants, a majority of whom were engineering professors, seemed to share the single basic conviction essential
to the vitality of Catholic higher education: that the Catholic intellectual tradition, and the Catholic Church, provides resources of mind and
spirit that can enrich the teaching and learning that define the mission
of the university and the vocation of its faculty and students. I have
always believed that, and it was refreshing to take part in an exchange
with first class scholars, Catholic and non-Catholic, who shared that
conviction.

In recent years a great deal of attention has been paid to the integrity of American Catholic colleges and universities, over two hundred
of them across the United States. Critics ask whether these institutions
are "really Catholic" because they seem to offer programs and courses
similar to those offered in other colleges and universities. Even warm
friends sometimes worry that one institution or another has gained
academic stature and fiscal stability but sacrificed its distinctive Catholic identity. In response, Catholic academic leaders point to first-rate
academic programs in theology, well-funded pastoral ministries, and
continuing commitment to a liberal arts core curriculum as evidence of
Catholic identity. And in recent years more and more institutions have
initiated faculty and staff development programs aimed at a wider

understanding of Catholic academic responsibilities. The Marianist University of Dayton has been a leader in this effort.

In 1990, after a long, world-wide consultation, Pope John Paul II issued an Apostolic Constitution on Catholic higher education, *Ex corde ecclesiae*. The pope strongly reaffirmed the church's commitment to scholarship and challenged Catholic institutions to assist the church by fostering dialogue between faith and culture. *Ex corde ecclesiae* also called for closer relationships between academic institutions and the hierarchy, and between Catholic theologians and the church's teaching authority. Tensions around those questions at times distracted attention from the overall commitment of the Catholic community, eloquently affirmed by the Holy See, to the work of Catholic scholarship and Catholic education. Translating that commitment into creative programs adapted to the ever changing landscapes of American society and American higher education is the task facing Catholic higher education today, and the Dayton conference demonstrated that some institutions and their faculties are meeting that challenge in the world of engineering research and education.

This conference was unique. To the best of my knowledge it was the first effort to bring together faculty and administrators who had been thinking about the relationship between Catholic commitment and engineering education. Of course many institutions, encouraged by *Ex corde ecclesiae*, have developed eloquent mission statements, new programs of faculty and staff orientation and development, curricular initiatives, usually centered on the core curriculum, and centers to provide space for study and reflection on mission-related questions. But all agree that the first priority is to convince the faculty that Catholic affiliation is a good thing, not just for the institution but for the practice of their vocation.

Among the most daunting challenges is the specialization of contemporary academic life, with faculty formed in discipline-centered graduate programs, conducting research on discipline specific, sometimes quite narrow issues. Faculty naturally think that they are not qualified to deal with questions of meaning and value, questions best left to the departments of philosophy, religious studies, or theology. Catholic and other church-related colleges and universities, altogether committed to serious scholarship, recognize the demands of disciplinary and professional integrity but cannot settle for segmentation that would leave the big questions of power, purpose, and personal and historic possibility to the humanities at school and to the church later on. Instead they search in their own way for the integration that has long been the promise of liberal education and the goal of Christian education.

In this setting the most impressive thing about the Dayton conference was the leadership, enthusiasm, and participation of a significant number of University of Dayton engineering faculty. Their commitment to the project and to exploring creative efforts to integrate their program into the Catholic mission of their own university shaped the spirit of the conference. I have visited many Catholic schools for mission-related programs, and I have rarely encountered faculty of a professional school so self-confident in discussing Catholic issues related to their area of responsibility. Several professors, Catholic and non-Catholic, explained that the source of their confidence was a faculty seminar facilitated by then-Provost and later University Professor of Faith and Culture James Heft, S.M. I knew that Dayton had over many years invested heavily in faculty development programs, encouraging faculty to find resources for their work as scholars and teachers in the Catholic intellectual tradition. Released time from teaching, stipends for summer seminars, research and writing, leadership from talented faculty, and support by administrators genuinely committed to supporting Catholic intellectual life on campus: all these have been provided at Dayton. One special seminar for engineers, well described in the proceedings, provided the backdrop for Dayton's own curricular initiatives and for the conference I attended. So here was evidence that investment in faculty development, conducted with intelligence and trust, can have genuinely constructive outcomes.

The conference demonstrated that Dayton is not alone. A number of other Catholic and Christian universities have taken steps to help engineering faculty confront mission-related questions. Several are reported in these papers. Almost all participants reported renewed appreciation of the liberal arts core curriculum, experiments in interdisciplinary and cross-disciplinary programs, especially with theology, and team teaching, all designed to insure that the humanities and social science courses genuinely contribute to a wider understanding of technology, design and related questions of human meaning and personal and social responsibility. Integrating core curriculum more fully into the engineering curriculum is no small trick, given the extraordinary demands of most engineering disciplines. But creative efforts are well described in these papers. Also evident in these papers is the tremendous resource of Catholic social teaching. Issues of poverty, gender, race, and environmental justice all relate to research agendas and public and corporate priorities, and Catholic social teaching provides rich resources for assessing these questions.

Finally, several participants not involved in Catholic higher education highlighted the way in which the initiatives taking place at Dayton and elsewhere reflect wider issues in engineering education and in contemporary culture. As these discussions took place I was reminded of theologian David Hollenbach's advice to those of us involved in Catholic higher education: beware of affirming our identity in terms of our difference and distance from others. Instead we should adopt a stance of intellectual solidarity. Solidarity draws us, in this case as engineers, to share the aspirations and the anxieties of the profession as our own. In seeking to bring Catholic resources to bear on engineering scholarship and teaching, one hopes to enrich the entire profession and, indeed, the surrounding society for which we also share responsibility.

This sense of solidarity is well expressed in the keynote address of John M. Staudenmaier, S.J., included as the preface to this volume. He affirmed the work reported at the conference and pointed directions for future conversations. Many papers at the conference approached the question from the inside, clarifying engineering method, focusing on the importance of design, exploring the meaning of the precision and discipline of the profession. Catholic artistic and humanistic traditions assist that exploration, while Catholic social teaching assists exploration of questions of meaning and value. Fr. Staudenmaier emphasized the context, the "fabric," within which this all takes place. Everything—everything Catholic as well as everything connected with teaching and learning, designing, and making—takes place within this world, which must be approached with affection, as he put it. In this sense the questions posed by faith, by the humanities, and by social and political responsibility are not extras, to be dealt with along the edges of education or practice, but inescapable and built-in, informing for better or worse the entire enterprise in which we are engaged. So Staudenmaier's call for first rate education in ethics and politics for all engineering students and graduate students (and thus by implication for all of us professors, staff members, and administrators) gives a sharp edge to the challenge of engineering education in the Catholic university. No one should think it will be easy.

I came away from this conference more encouraged about Catholic higher education than I have been in the years since I wrote a book on the subject. At this conference participants pushed the envelope a bit, examining the Catholic question with the same rigor they bring to their own disciplines. And they did so without the least suggestion of either defensiveness or triumphalism. For them this is not an experiment in denominational rehabilitation or religious competition but an effort to probe to the depths their own intellectual and vocational commitments. That is what the institution and culture of Catholic higher education is supposed to encourage and support. It was a privilege to take part. I hope I will be invited to the next gathering.

Worcester, Mass
January 10, 2006

Preface

Elegant Design Is Not Enough: Embracing the Tangled "We"
to Critique Technology

FATHER JOHN STAUDENMAIER, S.J.

There is a theological premise that underlies my thinking as I have prepared this essay on engineering and Roman Catholicism. It's a two-sentence description of what is distinctive about the Roman Catholic theological tradition (when contrasted with other Christian theological traditions). It starts like this: *The world is good more than it is suspect.* That is to say, the Roman Catholic theological tradition understands God to be approaching the fabric of reality as good, that God does not first of all approach the human condition with distaste or suspicion. The second sentence states: *Conversion is gradual and lifelong more than it is sudden.* Catholics baptize babies who are not capable of a life-changing decision (as in the expression "born again"). Catholic baptism, even for adults, is understood primarily as initiation into the believing community. The born-again adult baptism idea is rooted in a theological understanding that grace intervenes from somewhere outside your life and turns you around. Catholics do not deny such moments of dramatic conversion, but they tend to understand them in the context that conversion is gradual and lifelong. This implies that we should seek God primarily within the human fabric of our times, with the specific graces and temptations to which we citizens of this time and place are subject. God sends believers into the world but not from a starting point that is independent of the world in which the believers live. The starting point for being sent in the service of God's redeeming grace is my own need for redeeming grace, a need for conversion that lasts my whole life long. Thus, to conclude this introduction, I never approach my own human context as if I were not deeply embedded in it.

How does this theological approach bear on engineering education? Let me begin by suggesting that engineering schools need to teach their students to understand technologies according to two different standards for technological success. The first is this: a design succeeds if it works, given its constraints. I think all engineers understand this. And a truly elegant design, blending the realism of a project's constraints with engineering creativity, moves an engineer by its sheer beauty. The second meaning of success is less often talked about by engineers: a technology is successful when it gets embedded in its host society so deeply that if it were to suddenly disappear, society would be thrown into chaos. Consider two technologies, dental floss and paved roads. According to the first standard, dental floss is much more successful: it is elegantly designed for its purpose, economical to produce, effective, easy to dispose of. Roads are not—potholes, temperature-change damage, storm surge run-off flooding, and so on. But according to the second standard, dental floss is not very successful. If all the floss in the world disappeared, life would go on more or less as before. But if all the paved roads were to disappear, to be replaced with a network of rails for long distance travel and haulage and short-run roads subject to mud and dust or labor-intensive stone surfaces (the situation in the US ca. 1900) the United States could not function. Not today. Even if all the paved roads vanished, the layout of buildings, designed for car and truck traffic with large surrounding parking spaces, would be too far from rail heads to manage travel almost everywhere in the current US. You see the point. Take any deeply embedded technological system—water distribution, electric light and power, petrochemical fertilizers, pesticides and herbicides, the internet—and you find that, as it becomes more successful in the second sense, it also becomes so necessary a part of societal life that it ceases to look much like a technology in the first sense. Its design constraints become more or less invisible. It is just "the way things are." One only pays attention to a successful technology when something happens that calls attention to its constraints (as in wheelchair access to public buildings in the last quarter century).

Let's look at the two kinds of technological success in terms of engineering education. At the University of Detroit Mercy, we require a course titled The Politics and Ethics of Engineering for all our incoming undergraduates. I always begin the term telling students that "It is not my job to teach you to be technically competent. If you don't become technically smart, you will flunk out. But you could graduate from this engineering school being technically competent the first way and utterly incompetent the second way. You could go through this school and not take any of the tough professors in history, psychology, theology, philosophy, or literature. You could turn out as an engineer who can get a license to practice and create elegant designs and still be a fool because you don't understand the politics or ethics of technology. Should you graduate that way, you will practice engineering as someone else's tool; someone else will tell you what to do and define the constraints of your projects, and you, because you are ignorant of the fabric in which the technology operates, will end up following orders."

Engineers need to learn that, where money and power converge, technically smart people will encounter ethical and political pressure to deliver whatever it is that the people who are spending the money expect. But if you have failed to make a habit of interpreting the politics of your context and thinking through their ethical implications, you will not have positioned yourself to have a say about the larger policy questions that shape investment decisions. This is hard learning for engineering students and their faculty because the learning challenges for technological expertise are themselves so demanding. It is easy to be distracted from the demands of learning how to interpret the messy human condition.

If an engineering school is going to be worth its salt, however, it will embed into its undergraduate education a demanding, professional-level exploration of the second meaning of engineering success, in which technologies become so much a part of the normal life of society that their technical constraints become nearly invisible while they shape the lives and expectations of the citizens of that society. Is there not a moral obligation, in a society such as the United States that is dependent on extremely complex and interlocking technological systems, to debate investments in research and development and technological maintenance? If the technically expert people have not earned seats at the table where decisions are made—so that the techies tend to hang out drinking beer on the weekends with each other, talking about the fools who have set up these priorities that we're stuck with—then engineering education has failed. So goes the central thesis I am proposing.

To connect this analysis with a theological vision, let's suppose we have an ideal engineering school where all of the students have seriously studied history and philosophy and theology and literature, and you have another school that hasn't. Let us imagine ourselves as graduates from the technical school with little investment in understanding technological contexts, when we find ourselves concerned about what look to be dangerous and wrong-headed technological investment priorities about R&D and the maintenance of existing systems. On what grounds might we stand when we criticize technological investments in the United States? Do we stand as an elite group of disgruntled outsiders who look at all the jerks—the spin doctors, the politicians, the stupid consumers, the advertisers—from a metaphysically clean spot where we sit drinking our beer? We see stupid technological ineptness, but we have decoupled ourselves from the human fabric in the process.

Then the Jesus of Roman Catholicism comes in and drinks a beer with us and overhears our conversation. We can expect Jesus to say, "There's only one human fabric. There is no sanitized space from which to critique technological behavior or any other behavior. You're stuck. The people you're so upset with, they are you; you are part of this situation; you are not outside of it. I will not allow you to imagine that God sees the human condition as despicable or contemptible, that God is detached from it, looking at it as a toxic soup." If you're Roman Catholic, you're stuck with a God who never ever approaches the human condition—as stupid as it can be and as vile as it can be—with anything except the mysterious and profound affection that only God can create.

Now, the second mythical group is equally upset with the same event and drinking their beers too. But they're saying, "We have serious problems that we never even thought of." This second sort of engineer tends not to divide the world into an elite "we" and the technologically unwashed "they." They recognize that "we" are all embedded in the fabric that produced the political and societal context in which we live.

It's a tangled we, a fractured we. But the cost of my being part of this tradition is that I bear the sadness of being part of a "we" that I may think is in big trouble. I think Catholic engineering schools ought to find ways to teach their students that there's only one human race, not two—not the smart people and the fools, not the good people and the ignorant ones—that there's only one human condition. They do engineering recognizing that, when you are given the heavy burden of expertise, you are called by God to minister to yourselves, in terms of technological biases and expectations and omissions, as you minister to your fellow citizens of society. Catholic engineers need to learn how to use the word "we" in a reverent and respectful fashion. And the price of doing that: you have to trade in righteous contempt and be willing to embrace grief. The cost of loving the world is to be willing to rejoice in its beauty and to grieve over its burdens. And the subtle, seductive temptation for engineers—this is a challenge for any group of professionals—is to say "no, that's them, we are not responsible." What would it take for Catholic schools of engineering to embed in their curriculum processes by which engineers approach the tangle of the human condition with kinship and reverence and affection? What would it take to educate students to form a habit of self-awareness and a habit of understanding the nature of God's relationship with the human condition that is deeply embedded in the Catholic theological tradition? I think you start by looking at God and saying, "God does not find us contemptible, God is not put off by our violence, God works to heal our wounds with us, God chooses to love."

You must understand that God has a stake in the beauty of the human condition and as a result we have a calling. The better we get at this, the deeper the calling runs.

NOTE

This preface is adapted from my September 23, 2005, keynote address, at the conference "The Role of Engineering at a Catholic University," held at the University of Dayton. The original title of that address was "When I Critique a Technology, on What Ground Do I Stand? Successful Technologies from a Catholic Theological Perspective."

Acknowledgements

The works included in this volume were first presented at a national conference on "The Role of Engineering at a Catholic University," organized through the collaboration of members of the mechanical engineering and religious studies faculty at the University of Dayton. The conference, held at the university on September 22–24, 2005, attracted participants from ten Catholic universities, one Christian university, and three secular universities. The discussions were lively, the participation full, and, at the end, all agreed that there had not been in the past a forum for such important discussions.

Thanks are particularly in order for Dr. Margaret Pinnell, Dr. Brad Kallenberg, and Ms. Carol Wilbanks for their primary roles in planning and coordinating the conference.

Introduction

JAMES L. HEFT, S.M., AND KEVIN HALLINAN

Engineers are nearly invisible. Few people in the United States, even engineers, can name a single living engineer who enjoys a high profile in the national media. Not many people really know what engineers do. Surprisingly, some people still believe engineers are the people who operate trains, rather than the people who labor to invent and produce a huge number of the goods of the economy. By means of technology, engineers imagine, design, and construct all sorts of artifacts that affect not only the lives of individual people, but also the very environment in which they live, to say nothing of the environment the world over. Despite the transformative power of the work of engineers in the world today, the official teachings of the Catholic Church, except perhaps in regard to use of military weapons, have provided little in the way of critical analysis of the technology engineers create. Equally lacking in contemporary Catholic thought is insight into how technologies might best contribute to the common good. Given the ever-expanding scale of technology and its power over the global socio-economic situation, as well as the fragility of the earth's natural environment, this lack of attention to technology in Catholic thought needs to be changed.

1

Unfortunately, scholars in Catholic universities are to some degree complicit in this lacuna. While it is true that Catholic universities educate young men and women who enter a world dominated by the pervasive technologies created by engineers, they graduate few students with even a basic understanding of the role that technology will assume in their lives and culture. Moreover, the technical education that engineers receive at the sixteen United States Catholic universities offering engineering degrees largely mirrors the education secular universities provide their graduates; consequently, engineering graduates of both Catholic and secular universities all too often are similarly oblivious of the impact of the technologies they develop. Few of these graduates, even those of Catholic universities, have been challenged to think through the possible relationships of their faith to their practice as engineers.

We are unaware of any book that explores the role of engineering at Catholic universities from the perspectives of both humanities and engineering faculty, as well as their students. This book includes a rich variety of these perspectives, along with those of faculty from non-Catholic Christian universities who have also wrestled with how the Christian tradition should inform the engineering education they offer. It should be no surprise that the visions of engineering education that these faculty from Christian and Catholic universities share have much in common. It should also be recalled that Catholic universities in the United States require for all students, including engineers, courses in philosophy and theology. These requirements put faculty at Catholic universities in a unique position to consider the relevance of Catholic tradition to the development of technology. The chapters of this book, originally conference papers, provide initial but important steps toward two goals: first, helping Catholic engineering programs connect better to their founding traditions; and second, helping Christianity develop a vision for the appropriate development and use of technology globally.

The ten chapters that comprise this book have depended upon and embody in various forms collaboration between educators from multiple disciplines: theologians, historians, ethicists, engineers. A number of Catholic universities have invested substantial effort and resources to develop engineering faculty who are both interested in and capable of integrating a Catholic commitment to the common good with engineering education and research. Close collaboration between humanities and engineering faculty, though unusual, has contributed, as evidenced by many of the chapters of this book, to a distinctive and rich set of insights into an area far too long ignored: engineering and Catholic social teaching. Especially important was the contribution to this volume by engineers themselves—for in the end, it will be these faculty and their students who will need to understand and practice engineering within a Catholic vision of the person and society. Also important, however, is the substantive and sustained conversation between engineers on the one hand and philosophers, ethicists, and theologians on the other. Theologians, for example, who acquire a fuller understanding of the nature of the development of technology will be able to contribute to Catholic social teaching a body of knowledge now, as mentioned earlier, largely missing. Such cross-disciplinary work is clearly important, especially in the face of the following three questions that have preoccupied the thinking of the contributors to this text:

1. What does the Catholic faith and tradition, as well as other Christian traditions, have to say about the education of engineers and the role of engineers in society?
2. How has engineering and engineering education been influenced by Catholic social teaching?
3. How might a Catholic vision of the person and society influence more effectively both engineering and non-engineering curricula at Catholic universities to create truly distinctive graduates?

Obviously, none of these questions can be addressed by someone operating from within a single discipline, given the way most disciplines are compartmentalized today. A few theologians have some ideas about the role that the Catholic tradition should play in university education in general, but they will not likely understand the practice of engineering nor be able to relate adequately their expertise to professional education in general. Engineers for their part know best the practice of engineering, but need a greater understanding of the ramifications of the Catholic educational tradition for engineering. Moreover, engineers need the freedom and institutional support to imagine just how the relationship between the Catholic educational tradition and engineering practice might be embodied. In other words, engineers need theologians to broaden their horizons and theologians need engineers to concretize their thinking. Between engineering education and practice on the one hand and Christian social ethics on the other, we need a continuous *feedback loop*. This volume represents the first iteration in just such a process.

The foreword was written by historian and former Holy Cross Professor of Roman Catholic Studies (and now the University Professor of Faith and Culture at the University of Dayton), David J. O'Brien. He attended the entire University of Dayton conference on "The Role of Engineering at a Catholic University," September 22–24, 2005, at which many of the papers that eventually become chapters of this book were presented. In the foreword, he makes clear the importance of the integration of Catholic thought, especially Catholic social thought, with professional education. By the end of the conference, he felt, as he wrote above, "more encouraged about Catholic higher education than I have been in the years since I wrote a book on the subject" (referring to his widely read book, *From the Heart of the American Church: Catholic Higher Education and American Culture* [Orbis Books, 1994]).

In the preface, technological historian John Staudenmaier, S.J., explains two principles of Catholicism that should guide the education and work of engineers: first, the world is better than most people think, for God created it good and loves it; and second, conversion is a gradual process and growth takes time. As a consequence, engineers should consider critically and deeply the full consequences of *good* design, the sort of design that changes the fabric of society.

Part 1 focuses on the *shape and art* of engineering and the Catholic tradition, and the interplay between them. The image of shape and art rather than, for example, that of structure and practice conveys a sense that engineering and the Catholic tradition are not already defined. "Shape" suggests boundaries that are multi-faceted and evolving, whereas the word "structure" suggests boundaries that are clearly defined. "Art" conveys a sense of the importance of individual creativity through the use of imagination, experience, and interaction with others, whereas engineering as a structure and practice is often reduced to applied science. Such a definition, based on an understanding of engineering as precise and already defined, suggests that peoples' imagination, experience, and location in the world are irrelevant. Engineering as an art, however, suggests an understanding of engineering where things are not cut and dried, and where the personal insight and creativity of the practitioner play key roles. The nature of any problem, the system boundaries defining the problem, the resources imagined as being needed to solve the problem, and the inexactness of the "best" solution, all call into question any idea of engineering as an exact profession, requiring for successful practice an exact set of predefined skills. In a similar way, it would be a mistake to think of the Catholic tradition as fixed structures and preset practices. Such calcification of living tradition reduces it to endless repetition of past ideas and forms that admit of no change.

Part 1 also provides a picture of the shape and art of engineering and the Catholic tradition. In the first chapter, James Heft, S.M., describes an initiative at the University of Dayton that explored the shape and art of the Catholic intellectual tradition as it might relate to engineering education. Engineering and humanities faculty with reduced teaching loads explored together each week for an entire semester Catholic social teachings and the art of engineering. The faculty responses to the seminar reveal just how personally transformative such faculty development efforts can be. In chapter 2, Brad Kallenberg, a theologian and a teacher of engineering ethics at the University of Dayton, explores the historical foundations of engineering. In the end, he concludes that key objectives of contemporary engineering can be traced in large part to medieval monasticism. The advantage of having this longer historical perspective is found in the clues it provides for the moral practice of engineering that are more substantive than one typically finds in the available accounts of engineering history. Specifically, he suggests that the medieval theological underpinnings of engineering present a picture of artifacts that could be considered inherently sacramental in that those artifacts are understood as visible human creations, the results of mechanical reasoning, that reflect the creativity of God.

Part 2 offers examples of how in practice the bridge between Catholic tradition and engineering has been crossed from both the engineering side and that of Catholic traditions, and, more importantly, how traffic on this bridge might be increased in the future. At the University of Dayton and Santa Clara University, where a great deal of thinking about the role of Catholic tradition in engineering education has taken place, we see two factors that have increased that important traffic: (1) engineering faculty development relative to Catholic tradition; and (2) faculty development aimed at promoting dialogue between faculty in engineering and the humanities. At the University of Dayton, one vehicle for achieving this flow has been through year long multi-disciplinary faculty seminars, including the one on Religion and Ethics in the Professions, out of which the 2005 conference grew. At Santa Clara University, the Ignatian Center for Jesuit Education has been instrumental in helping the university community realize the evolving role and implications of their Jesuit identity and faith tradition as they express themselves in works of justice. Further, their Center for Science, Technology, and Society, which has as its mission the illumination of the interplay of science and technology with culture and society, has promoted dialogue between the humanities and engineering.

As described in the leading chapters of this second part, by Kevin Hallinan and Margaret Pinnell of the University of Dayton and the then Dean Daniel Pitt of Santa Clara University, the engineering curricula and humanities, arts, and social sciences curricula have been affected by these faculty development efforts. At the University of Dayton, the Catholic influence has been felt in a number of ways. First, sustainable engineering practices, driven by recognition of the sacramentality of creation, have been integrated into much of the engineering curriculum. Second, in a desire to serve the poor, students have been engaged in meaningful service-learning opportunities, both local and international, primarily through a student-designed organization called ETHOS (Engineers in Technical Humanitarian Opportunities for Service-Learning). Third, there is a strong emphasis on creating a community among students and faculty. At Santa Clara University, the formation of engineering students takes place especially through community-based projects that demonstrate the social relevance of engineering and educates students of conscience who show compassion to those who need it. Pitt attributes this focus to the university's ability to draw into their programs women and under-represented minorities from the socially and economically diverse San Jose area in which they are located.

All Christian engineering programs emphasize education of the "whole" person; thus, learning in the humanities (including theology or religious studies and philosophy, which are viewed as being at the "heart" of Catholic universities), arts, and social sciences is deemed vital and important. Included in this integrated vision of education is engineering ethics from a distinctly Christian perspective. Dialogue between engineering and theology and philosophy faculty has led to the creation of some unique engineering ethics courses. For example, Hamid Rafizadeh and Brad Kallenberg advocate in chapter 5 a systems view of time-dependent ethical decisions in engineering environments. Through dialogue with engineers, they have learned the importance of bringing ethics into the very beginning of a design process. Ethical goals, they argue, should be included in a portfolio of goals for any design.

In chapter 6, Marquette theologians Jame Schaefer and Paul Heidebrecht likewise describe a novel graduate seminar course aimed at discerning and demonstrating ways of "constructively relating theology and technology." Their seminar underscored the paucity of Catholic thinking regarding technology. Among the numerous benefits of their seminar was the recognition that Catholic theology's relationship to technology can be both affirming and yet critical, especially when it comes to those technologies that violate life.

Building on the general community-based service-learning element of a Catholic engineering education described in part 2, part 3 provides two detailed examples of how Catholic social teaching informed international service-learning experiences. In chapter 7, Camille George, an engineer, and Professor Barbara Sain, a theologian, both on the faculty of St. Thomas University, use the theme of integrated human development to inform engineering practice. They horizontally integrated an international service-learning design project course and an elective upper-level theology course entitled Christian Faith and the Engineering Profession in ways that gave students a Catholic framework for evaluating actual engineering projects designed to promote human development. In chapter 8, Marquette engineering faculty members Daniel Zitomer, Lars Olson, and John Borg describe their international service-learning program. Of particular interest is the use they make of the Ignatian concept of "indifference" that, for example, teaches civil engineering students to adapt and make field modifications according to the actual needs of the people they are helping rather than according to their own preconceived ideas about design.

Part 4, the final section of this book, provides still more detailed examples regarding another connecting theme between Catholic tradition and engineering education that part 2 painted with a broad brush, namely, the formation and preparation of students so that they see their profession as a vocation. Most important in this formation process is helping students to reflect at some length on the "why" of their profession. In chapter 9, University of Dayton engineering technology professor Scott Schneider developed a national survey to learn what role thinking about engineering as a vocation played in mostly religiously-affiliated universities. He found that more than half of the faculty respondents worked to foster a vocational awareness in their students. That awareness was primarily developed implicitly through personal interactions with students and by faculty example, and explicitly through engineering volunteerism and ethics courses. Perhaps the best example of formation is seen in chapter 10, contributed from Valparaiso University engineering faculty members Carmine Polito, Douglas Tougaw, and Kraig J. Olejniczak. Throughout their curriculum, they present a vision of engineering as a career through which their graduates can make the world a better place. Students are continuously asked to dedicate their lives to helping others in their work as engineers, an emphasis that naturally leads to discussions about engineering ethics, environmental stewardship, and social justice. Faculty typically rely on service learning, both curricular and extracurricular, to help students think of their profession as a vocation.

In conclusion, this book makes clear the profound need for much more work on how Catholic intellectual traditions, and more particularly Catholic social teachings, can address in an informed way professional education, especially engineering education. This need was already anticipated in Pope John Paul II's *Apostolic Constitution on Catholic Universities: Ex corde ecclesiae,* published in 1990. In paragraph seven of the constitution's introduction, the pope wrote:

> In the world today, characterized by such rapid developments in science and technology, the tasks of a Catholic university assume an ever greater importance and urgency. Scientific and technological discoveries create an enormous economic and industrial growth, but they also inescapably require the correspondingly necessary *search for meaning* in order to guarantee that the new discoveries be used for the authentic good of individuals and human society as a whole. If it is the responsibility of every university to teach for such meaning, a Catholic university is called in a particular way to respond to this need: Its Christian inspiration enables it to include the moral, spiritual and religious dimension in its research, and to evaluate the attainments of science and technology in the perspective of the totality of the human person.

This volume offers a variety of examples of how that work is being done, and how more of it needs to be done. The thought of faculty and students at Catholic and Christian universities contained in this volume contributes, we hope, to that necessary *search for meaning* mentioned by the pope. We live in a culture deeply shaped by technology. The needed dialogue among theologians and engineers and philosophers and still others can be both rich and demanding, but above all, that dialogue is needed more now than ever.

PART 1

The Shape and Art of Engineering and the Catholic Tradition

Chapter One

Exploring a Catholic Vision of Engineering

JAMES L. HEFT, S.M.

For several years now, the University of Dayton has been making substantial investments in faculty development. Some of those investments are especially focused on relating the research of the faculty to a curriculum that expresses the intellectual dimensions of the university's founding religious tradition, Catholicism. Several ways of doing this include research grants, endowed chairs held for four years by prominent members of the faculty, reading groups that meet occasionally or even for a full academic year, and cross-disciplinary faculty seminars. This paper describes one of those cross-disciplinary faculty seminars, the fourth in a series of such seminars,[1] this time focused on Engineering and a Catholic University.

Planning for such a seminar requires support from the Provost's office, the dean of the School of Engineering as well as other deans (in this particular seminar, the dean of the College of Arts and Sciences), and consulting the chairs of the departments from which we hoped to invite participants. We look for tenured younger faculty who have demonstrated academic excellence in their discipline and leadership with their peers. We also invited some very experienced individuals

13

who can speak firsthand, in ways helpful to faculty and their research, about the mission of the university. Finally, in this particular seminar, we tried to include faculty from disciplines besides engineering who can contribute fruitfully to the cross-disciplinary conversation.

The engineering seminar, which in January of 2005 began to meet for two hours weekly, included twelve faculty members, four from the school of engineering, along with a historian of science, a philosopher with a graduate degree in computing who teaches engineering ethics, a mathematician, a professor of physics, a theologian who teaches engineering ethics, the former president of the university and himself an electrical engineer, the head of the university's Research Institute (which carried out over $65 million worth of sponsored research that year), and me, a historian and theologian. We found it necessary to invite already in August 2004 the faculty we wanted to participate in the seminar so that the January 2005 teaching schedules could be coordinated in such a way as to make a weekly meeting of all the participants possible. Each participant received a reduced teaching load for that January term along with a $5,000 summer research stipend. The financial support for this initiative drew upon funds from the provost's office, the deans of the School of Engineering and the College of Arts and Sciences, as well as an external grant won through the author's efforts. During the fall of 2004, I met regularly with four members of the seminar—two engineers, a historian of science, and a theologian who taught engineering ethics—to discuss readings for the seminar. In the process of planning the readings and sequence of topics, this group regularly consulted all the other members of the seminar to insure their understanding and support of the proposed semester's syllabus (see Appendix A). These readings of the syllabus explored three themes: the history of Catholic higher education in the United States, the history of the profession of engineering, and, finally, Catholic intellectual traditions, especially Catholic social teachings and Marianist approaches to education.[2]

WHY THE THREE THEMES?

Before discussing some of the highlights of the seminar, it might first be helpful to say something about the three themes that formed the core of the syllabus of readings: the history of Catholic higher education, the history of engineering, and Catholic intellectual traditions. One sixth of all Catholic colleges and universities in the world are in the United States, and among them some of the very best Catholic colleges and universities in the world. This extensive array of 230 or so quite diverse institutions has been established by only 6 percent of the world's Catholics. Most of them were founded as an alternative to public education, which in the nineteenth century was mainly Protestant, and in the latter part of the twentieth century mainly secular. Over 90 percent of these Catholic institutions were founded by religious orders, almost all of which had been suffering a loss of membership since 1965, when the Second Vatican Council ended. By that time, many Catholics had already exited Catholic neighborhoods and schools and moved up the social and economic ladder and out to the suburbs. Now Protestants are no longer the enemy but rather separated brethren and valued colleagues, and many secular institutions have superb academic offerings and active Catholic campus ministry programs. What all these profound changes have meant is that Catholic colleges and universities now need to find a different rationale, different and I believe deeper, than the one that distinguished them in the nineteenth and first half of the twentieth centuries.

And besides these changes external to Catholic colleges and universities, internally we now have lay leaders and faculty members who are more professional in their disciplines and pluralistic in their beliefs than ever before, and students who also now come from more diverse backgrounds, some of whom are nontraditionally aged students. In other words, there is a greater need now than ever before to rethink the founding purpose of these institutions given the changed environment in which they exist and the different internal constituencies they embody. Today, many faculty, but not all, do not have the skills to blend professional and technical competence with moral and religious reflection. Given the explosion of engineering knowledge in recent decades and the increased expectations of what students are expected to know upon graduation, faculty have created engineering curricula that focus almost exclusively on technical issues. As a consequence, many students do not have the religious and moral literacy needed to sort through more complex societal issues in which they are expected to function as engineers who will make a difference. Without greater clarity of purpose for professional education, Catholic colleges and universities, deeply affected by these many and far-reaching changes, are likely to become indistinguishable in their curricular requirements from their secular counterparts. While Catholic institutions would often benefit in many ways if they emulated the curricula and research done at many secular institutions, they also, in my view, need to ask themselves how being Catholic might make their curricular distinctiveness a valuable contribution to the pluralism of professional higher education in this country.

The second theme of the seminar was the exploration of the history of the profession of engineering. Most faculty are notoriously "presentist"; that is, most faculty are understandably doing research on new and developing ideas and events. Even some historians are oddly unaware of how history was written centuries—even millennia—ago. The group of seminar planners believed that a study of the history of engineering would show that the way in which engineering is done now is not the way that it was done a century ago; this also implies that the way it is done now is not the only way it can be done, and the imagination, enlightened by history, can begin to imagine still other ways that engineering might and should be done in the future. The degree, for example, to which government priorities funded engineering research at universities following World War II is extraordinary. When the participants read some of the reports on the purposes and shape of engineering education written over the past century by national review boards and accrediting boards, they saw a fairly wide range of positions that sometimes stressed the theoretical over the practical. They also saw, and still experience, the difficulties professional schools have meeting the competing priorities of general education, modern engineering theory, and a more hands-on approach to engineering. Understanding this history makes it easier to think about new ways engineering education might be developed. One member of the seminar found this part of the seminar so valuable that he suggested that we draw up lesson plans on the history of engineering education.

And finally, the theme of Catholic intellectual traditions is significant for a variety of reasons. First, few people, including cradle Catholics and academics who are devout Catholics, understand that their religious tradition embodies a long and multifaceted intellectual tradition that includes not only philosophy and theology, but aesthetics and art, music and drama, architecture and jurisprudence, history and literature, and carefully developed ideas of the nature of the human community and how it is best to live together. More specifically, especially over the last century or so, the Catholic Church has produced a body of what is called "social teaching," the most recent additions to which emphasize the environment, issues related to international development and social justice (with particular focus on war, economic issues, and the needs of the poor), and the best uses of technology. One of the purposes of the cross-disciplinary seminar was to acquaint faculty with those traditions that might enrich the engineering curriculum.

That body of Catholic social teaching has been developed through a method of reasoning not so distant from the way engineers size up and tackle problems. Already with the Greeks, we have descriptions of "practical reason" that involved three steps: first, the recognition and definition of a problem, resulting in an understanding of the gap between "what is" and "what ought to be"; second, a formulation of different ways or options (designs) to close the gap, and a process on how to choose the best way; and third, the formulation of steps to implement the design that is chosen. In this sense, it can be said that professional knowledge is a tradition, that is, a historically extended and socially embodied network of arguments and practices that arrive at desired solutions, or at least improved situations. Catholic social teaching is also a tradition of such practical reasoning that focuses on important social questions. Catholic social teaching should therefore be a natural dialogue partner with engineers interested in extending engineering methods to include more explicitly social, economic and ecological consequences of engineering designs. We are only beginning to learn the possible relationships between these two forms of practical reasoning.

SOME HIGHLIGHTS FROM THE SEMINAR

While it would be possible to report the content and describe the tone of the conversations of twelve faculty who met for two hours every week over a period of nearly four full months, it will be more useful, I believe, to concentrate instead upon some of the most important moments in the seminar.[3] We began with a full-day meeting in mid-January. The better part of that day was set aside to introduce the seminar to all the participants, and them to each other. Faculty seem no different than most people who, if given the invitation in the right atmosphere, will talk at some length about themselves and their hopes and convictions, as well as their hesitations and the challenges they face. In fact, one of the most important outcomes of this seminar is the friendships that were developed between faculty who otherwise might not even have met, much less gotten to know each other. It took some time to establish common expectations, which was not easy when people come from quite different disciplines. Even with the same profession—engineering—one member expressed concern that he saw few signs of Catholic or Marianist influence in the School of Engineering, while another expressed reservations about the expectation that engineering professors should help form students in their faith. Despite such differences, some of the participants have gone on to collaborate on projects, and two went on to team teach a new course. It is commonly observed that academics typically come together only to complain about parking problems and strained budgets. A seminar like this, however, supports prolonged conversation about topics more significant than parking and even budgets, and allows time and space for participants to teach their colleagues about that to which they devote their time and energy.

One of the significant continuing seminar conversations focused on history, not only on the work of engineers in the past and the goals of engineering education then, but also on the history of Catholic higher education. The work of engineers in the Middle Ages and their sense of "vocation," the professionalization of the disciplines at the end of the nineteenth century, the key role played by military needs, and the amount of government funding after World War II that poured into many universities with engineering schools—all these historical developments helped clarify the important ways that forces external to the university (such as culture and government priorities) shaped and continue to shape engineering education. Historical perspectives also illuminate some of the pretensions and overblown expectations from which some of our predecessors suffered. For example, shortly after World War I, Henry Gannt, a management consultant, announced confidently that the leaders of American society should no longer be politicians or financiers, but engineers: "opinions must give way to facts, and words to deeds, and the engineer, who is a man of few opinions and many facts, few words and many deeds, should be accorded the leadership . . . in our economic system."[4] In 1930, the fiftieth anniversary of the American Society of Mechanical Engineers, a play was commissioned to demonstrate the many triumphs of American engineering; when that play was performed, those triumphs were portrayed as allegorical figures who, dressed in classical Roman togas, represented intelligence, imagination, mature control, conversion, and finance.[5] We might tend to chuckle at such a portrayal, but it was performed and presumably received as appropriate in its time. The more sober analysis of the contribution of engineers following events such as the dropping of the Atom bomb and the space shuttle disaster now makes such confidence in the possibilities of engineering more modest. Then again, in the aftermath of Hurricane Katrina, one only wonders why the predictions and warnings of civil engineers about the vulnerability of New Orleans were not taken more seriously years before.

Engineering professors distinguish teaching theoretical dimensions of engineering from teaching students how to perceive an engineering problem accurately and then how to produce a creative solution. The best professors try to teach students to create imaginative designs. Such skills, however, are most likely learned by observing an engineer who practices the art, much like medical students learn the art of medicine by observing a clinical professor making rounds and interacting with patients. Studying the famous Ideo design process helped the seminar participants understand the difference between engineering as a "science" and as an "art."[6] It is, I believe, especially helpful for non-engineering faculty like myself to realize that engineering education is about more than measurements and mechanics. Engineering makes regular use of mathematics and some science, but engineering at its best is also an art, even an aesthetic. The engineering professors in the seminar estimated that a relatively small percentage of their engineering students strive to master not just the science but also the art of engineering.

The history of Catholic higher education in the United States underscores its original purposes in an earlier age, as well as its contemporary context and accompanying challenges. That for most of their histories Catholic colleges and universities in the United States could count on religious brothers and priests and sisters to lead them and provide a good deal of the classroom education simply meant that the question of whether these institutions were Catholic or not rarely if ever arose. At the University of Dayton, we had brothers with European Ph.D.s in the sciences and engineering who founded and led these departments for decades. They were distinguished teachers and scholars, and many alumni attest to the profound impact these religious men had on them. Now they are gone, and one of the questions we face is whether we can and should communicate that distinctive form of education in our new context. I believe that there is enough concern today about the purpose of engineering education and about the mission of Catholic universities to create together a call for some creative and deep thinking about how engineering might be done distinctively in a Catholic university.

One of the guiding questions in the seminar was whether the curriculum of the School of Engineering should be different because the University of Dayton is a Catholic university. No one is so naïve as to think that there might be a special Catholic insight into thermodynamics or a Marianist take on hydraulics. Statics is statics, whether you are talking about a cathedral or the world headquarters of National Cash Register. On the other hand, we continued to press the question of how the curriculum might be distinct. To help create an approach to the question about a distinctive curriculum at a Catholic university, we used a three column exercise (see Appendix B). The first column identified key theological insights on which the University of Dayton is founded. The second column "translated" those insights into educational emphases. The third column, which was not filled out, left space for how those theological insights and educational emphases might appropriately appear in the curriculum of the School of Engineering.

We had to have several serious conversations on this matter before we were able to come up with some concrete suggestions. One member of the group made a combined list of all the suggestions to write into the third column. I believe that we may have made a mistake by not pressing for further discussion on the third column. Nevertheless, seeds were planted, and some of the research and course development that came out of the seminar sprouted some useful ideas for that column.

The group took an important step forward when we examined the most recent ABET (Accreditation Board of Engineering and Technology) "Criteria for Accrediting Engineering Programs." Under criterion 3, the following two outcomes are listed: "an understanding of professional and ethical responsibility" and "a broad education necessary to understand the impact of engineering solutions in a global and societal context." Also, criterion 4 includes a requirement that students typically be exposed to economic, environmental, ethical, health, and social issues—all as a part of a major design experience. All members of the seminar agreed that such issues should be included in engineering analyses and design. However, some of the engineers expressed hesitation about teaching such issues because it would be hard to measure their outcomes. Others disagreed, pointing out that students could be evaluated on how well they did on analyzing economic, social, and cultural consequences of various designs in much the same way other professors analyze papers in philosophy and theology—on whether, for example, they made thoughtful and coherent arguments. Further study needs to be done on how to make best use of the open-ended goals ABET identifies for such issues, especially those that touch directly on Catholic social teaching. Much also needs to be said about Catholic social teaching, but it cannot be fully covered in this essay. It is important to remember, however, that reaching such goals requires a constantly evolving body of practical reasoning that articulates the social consequences of a theological vision for the well-being and flourishing of the human community, both local and global.

Two further conversations in particular generated a high degree of energy. One of them focused on Bill Koen's book *Discussion of the Method* (Oxford University Press, 2003). We read only the first half of the book (pages 1–110), which focused on the heuristics of the engineering method. I found the most beneficial aspect of Koen's presentation to be how he made clear the way in which engineers can and ought to proceed in solving problems in the midst of ever-present but often changing constraints. Koen's sophisticated descriptions of the differences both between science and engineering and between an ideal solution and what was possible, given the "state of the art" (sota), I also found helpful. Some members of the seminar had serious reservations about Koen's thesis. They judged it to be arrogant in that they believed that he assumed that the method of the engineer could be used in all fields, not just engineering. The universal use of the Koen method presumes, they argued, cultural relativism and excludes moral considerations as irrelevant to the work of design and problem solving. At one point, Koen writes, "the engineering model is not based on an eternal or absolute value system, but on the one thought to represent a specific society. In a society of cannibals, the engineer will try to design the most efficient kettle" (19). Others disagreed, arguing that Koen would include morality, perhaps by inserting in the design process an ethical sota requirement. They also argued that Koen would not advise engineers to give society whatever it may want. Still others in the seminar pointed out that since emotion and narrative play a role in how we perceive and frame problems, then the engineering heuristic cannot be limited to techniques of efficiency alone. Narrative rationality is also important. As a description of practical reasoning aimed to solve practical problems, Koen's method seems appropriate. But not everything in life is a problem to be solved; some things in life are realities to be appreciated.

The second conversation that generated considerable energy had to do with the relationship between the mission of the university as Catholic and the amount of Department of Defense research that the university's Research Institute and engineering faculty do. When this issue was raised, we studied some documented history of earlier conversations at the University of Dayton on this matter. Those documents included the US bishops' 1983 pastoral on war and peacemaking, a report done by a committee of university faculty in 1985 evaluating the bishops' pastoral and the university's research, and finally a 1986 collection of eight short essays by university faculty on the morality of doing research for then President Reagan's Strategic Defense Initiative. It was not easy for some participants actually involved in military research to get past being defensive about the subject. Nor was it that helpful simply to assert that Catholicism is not a pacifist Christian tradition. Statements that "we have a right to defend our way of life" raised questions not only about the morality of the Iraq war, but also about whether we all would want to defend the consumerism, affluence, and the immorality celebrated in the media and available on the internet. We need, it would seem, to find better ways to discuss the appropriateness of research that supports legitimate forms of self-defense and distinguish it from the type of research that directly supports those bent on the United States exercising global control. This difficult conversation is the kind of conversation that a Catholic university should be able to take up and carry on honestly and openly. During our seminar, we made a good start, but it also became clear that we have some distance to go.

One final area that proved to be of considerable interest was the issue of gender and engineering. Women are entering the professions of law and medicine at a much higher rate than they are entering engineering, especially mechanical and civil engineering. It has been very difficult for the University of Dayton and other universities to hire and keep engineering faculty who are women. Besides the difficult and not yet complete adjustment that higher education, not that many years ago a mainly male preserve, needs to make to welcome both men and women, there is also the question of natural aptitude, workplace/discipline hospitality, and socialization that make the question of women in engineering both interesting and volatile, as the president of Harvard, Larry Summers discovered when he made comments about women in engineering and the sciences.[7] Anecdotal information is available, but serious study of the actual situation that affects women in the academy is only now being undertaken across the country. Gender and engineering is a difficult and complex issue, one that we need to face in a more open and honest way. Some members of the seminar have decided to devote their summer research to this issue.

THREE EVALUATIONS BY SEMINAR PARTICIPANTS

These intensive cross-disciplinary faculty seminars offer valuable opportunities to transcend the stereotypes we too often impose on colleagues in other disciplines, especially those disciplines that are quite different from the ones we ourselves practice. Bridges of understanding are crossed when engineers learn that some of their colleagues in the humanities are fascinated with technology, and when some of the humanities folks learn that some of the engineers think deeply about the nature and social value of their work. On several occasions, members of the seminar who were scientists or mathematicians offered ways for engineers and humanists to understand each other better. Moreover, the kind of seminar research projects (see also Appendix C) that the participants worked on will only enhance their appreciation of the cross-disciplinary challenges and opportunities that engineering poses within a Catholic university. By the end of the seminar, for example, all members of the seminar thought that an introduction to engineering and technology should be a required part of the general education curriculum. Given the pervasive impact technology has on our society, the need for such a course should be obvious. If such a course were taught in a Catholic university, the relevant issues of Catholic social teaching should be an integral part of it.

These conclusions are reflected in the comments of three of the participants in the seminar. They illustrate the importance and value of faculty development related explicitly to the mission of Catholic education. Even a religiously skeptical member of the engineering faculty, an engineer who initially expressed concern with his place in the seminar as an atheist at a Catholic university, said at the end of the seminar that the experience helped him appreciate the values emerging from the Catholic and Marianist heritage. He notes that applications of these values create opportunities for building a novel "connected" curriculum. Another engineering faculty, a convert to Catholicism, recalled that at the beginning of the seminar he mistakenly thought that religious influence in the School of Engineering was nonexistent. After participating in the seminar, however, he acknowledged surprise at the extent to which his engineering colleagues do in fact participate in the mission. Finally, a faculty member from the philosophy department, who had until a few years ago spent most of her professional life in non-Catholic universities, explains that her participation in the seminar was mostly based upon her interest in exploring unique ways that

Catholic universities can address gender equity issues, especially in the School of Engineering. Through interactions during the seminar with engineering faculty, she acquired a much greater respect for the environment in the School as women experience it. In this regard, the seminar and her research helped erase some stereotypes.

Bradley Duncan, Associate Professor, Electrical and
Computer Engineering

I suspect that I was asked to participate in Fr. Heft's seminar because I have often lamented—sometimes in public—that I don't really see evidence of the Catholic and Marianist nature of our university reflected in the day-to-day practices of the School of Engineering—and I've been here for more than fourteen years. To be fair, however, I must admit that I am a rather solitary person by nature. My failure to have experienced the Catholic nature of the School of Engineering may well be due—at least in part—to my failure to have searched for it.

So, I was quite excited to be asked to join Fr. Heft and the other participants last winter to discuss the nature of an engineering school at a Catholic university. I looked forward to discussing the history of the University of Dayton, especially as it would relate to our Catholic and Marianist educational heritage—which we did. I looked forward to discussing the unique characteristics of Catholic higher education in general (we did), and how these characteristics have come to influence the practices of schools of engineering at other Catholic universities. I looked forward to learning of unique practices within the School of Engineering that I could point to when parents of prospective students sit in my office, as they often do, and ask me just exactly what it is that's Catholic about the university and the engineering education their child will receive. I also very much looked forward to brainstorming new and creative ideas by which we in the School of Engineering could better live out and live up to our Catholic and Marianist heritage and identity. For the most part, we didn't do these latter things.

However, the seminar was not about satisfying my needs. The needs and wants of the collective pulled our discussions in other, equally interesting and equally important directions. We learned, for example, much about the history of engineering in general, and the nature of engineering design as the often creative interface between science and art. We had some interesting guests, including Peter Stein-fels and Mark Schwehn, and had plenty of illuminating readings to study, including Billy Koen's book *Discussion of the Method*. Most importantly, this seminar provided me with the opportunity to meet and talk with colleagues from "the other side of campus." Developing these new friendships was, for me, the most rewarding aspect of our seminar. We engineers (I, at least) tend to live in our own little worlds. It was refreshing to have the opportunity and venue to reach out and expand my collegial relationships. That was a very big deal for me, indeed.

Regrettably we spent only a little time struggling with questions related to how we might more fully integrate Catholic and Marianist ideals into the principles and practices of the School of Engineering. Why? Well, first of all these questions are really tough to answer, and our time was limited. Secondly, I have perceived, perhaps incorrectly so, that there may be some resistance to tackling these questions—a trait that is not unique to seminar participants. By the way, I may in fact be projecting onto the seminar participants my own take on comments made by a few engineering faculty over the past several months. Never-theless, I have sensed a hesitation based perhaps on a fear that by an-swering these questions, the School of Engineering might end up being too Catholic and thereby potentially less attractive to faculty as a place of employment for those who are non-Catholic. Yet, these are the very questions that *still* interest me the most. All these questions lead me to a discussion of the research I selected to emerge from my participation in the seminar.

I personally find it unsatisfying simply to look over our collective shoulders and point out that this or that practice or this or that action was *in keeping* with our Catholic and Marianist nature. One of my goals, then, has been to work with Joe Saliba, the dean of the School of Engineering, on ways *by design* to live out our Catholic and Marianist mission in a systemic and forward thinking way. I am not suggesting, however, that we somehow find a way to provide our students and guests with some sort of Catholic liturgical experience when they enter our engineering labs. I also think that more than just the curriculum should be influenced by our Catholic and Marianist heritage and traditions. (Actually, I think the curriculum may change little, if any. There won't be Catholic circuit analysis, for example.) However, looking into how the school's policies, practices, hiring decisions, advising, student recruitment, community service, and so on might be influenced in some unique way by our Catholic and Marianist heritage could be of great value. Our core focus will always be on generating outstanding engineering graduates. Our constituents will insist upon this. However, the framework in which we accomplish this goal can maybe be uniquely tailored and influenced by our special religious heritage.

There are sixteen Catholic universities that have schools of engineering; the University of Dayton is the only Marianist University with a school of engineering. We are thus in a position to offer a truly unique educational experience to our students if we accept the challenge to embrace our heritage and traditions fully. It is my belief that responding to these challenges will assist us with reputation building by helping us develop genuine distinctiveness in our policies, procedures, processes, and maybe even curricula by virtue of our Catholic and Marianist heritage. It will also help us by providing talking points for use in recruitment and advertising. All of this, of course, will have the ultimate benefit of helping us produce truly distinctive graduates.

So, with the assistance of Dr. Bill Moroney of our Psychology Department, I have developed a survey to help assess the Catholic and Marianist character of the School of Engineering. This first survey is intended for faculty and staff only, though if the survey is received well and produces promising results, additional surveys may be developed for students and alumni. The intent of this survey is to gauge how well we respond to and live up to the five characteristics of Marianist universities, specifically as they have been adapted for our work in the School of Engineering as part of the school's top-level strategic plan. In the survey I ask both "do we" and "should we" questions for each characteristic. I also ask for examples of how faculty and staff have seen these characteristics brought to life in the School of Engineering, and I also ask for suggestions for improvement.

For reference, the five characteristics of Marianist universities described in the 1999 working paper "Characteristics of Marianist Universities" are:

> Marianist universities educate for formation in faith.
> Marianist universities provide an excellent education.
> Marianist universities educate in a family spirit.
> Marianist universities educate for service, justice and peace.
> Marianist universities educate for adaptation and change.

My survey has been developed in the ABET spirit of continuous quality improvement. For example, each of the five characteristics in essence embodies an experience our students will have prior to graduation. Similar to ABET "outcomes," these characteristics should at least be demonstrable, if not directly measurable. This first survey will provide us with a baseline assessment, and I am hopeful that we will be able to identify several "high yield" initiatives (low cost, easy to implement, with wide impact and high visibility) for improvement.

The survey results have been striking. Faculty care about the Catholic and Marianist mission. Almost all say that it is important to them. While such a finding does not necessarily translate to a curriculum infused with "mission," it does evoke an idea that the ground is fertile for continued faculty development. In my mind, this environment is exciting.

Peggy DesAutels, Associate Professor of Philosophy

Ordinarily, faculty get together with other faculty for committee meetings and not much else. This seminar, however, allowed faculty from a number of disciplines to discuss interesting readings, explore new ideas, and develop new strategies in relation to engineering at Catholic universities. What an intellectual treat! I was the philosopher in the group with a graduate degree in computing who teaches engineering ethics. My research focuses on feminist philosophy and moral theory, so I was especially interested in exploring unique ways that Catholic universities can address gender equity issues in the School of Engineering. I was also interested in pursuing discussions on the ethics of conducting engineering research at Catholic universities that advances our country's war-related goals.

As Jim Heft has already noted, a number of interesting faculty projects came out of this seminar. These projects were generously supported by summer grants. In my case, I teamed up with Daniel Farhey, a civil engineer, to begin assessing and addressing the climate for women in the School of Engineering. Professor Farhey began work on a questionnaire that would help us to assess more quantitatively the current climate for women undergraduates in the School of Engineering at the University of Dayton. I took a more qualitative approach and formed a focus group comprised of seven women from three different departments in the School of Engineering. Participants in the group were asked to share and reflect on their backgrounds, interests, and experiences in the School of Engineering at the University of Dayton. To encourage an honest and open conversation, participants were assured that the individual sources of particular comments would remain confidential. Participants were also told that their contributions would be summarized in a report that would be shared with others at the university.

Some of the highlights of this focus group included their future family-related plans and their comments on the climate for women in the School of Engineering. Not a single participant expected that after completing her schooling she would pursue an uninterrupted, full-time career in an engineering-related job. Only one expected to pursue graduate work. One expected to be a stay-at-home mom, and the remainder expected either to cut back to part-time or to take several years off in order to raise a family. It would be interesting to explore further whether women engineering students at Catholic universities are more focused on family than women at other universities and, if so, to what degree this focus prevents them from pursuing engineering as a full-time career. It would also be interesting to explore to what degree the family-like atmosphere at Dayton and other similar Catholic universities could contribute to recruiting and retaining women in engineering.

All participants felt positively about the overall climate for women at the University of Dayton's School of Engineering, and most reported little to no direct harassment or sex discrimination. That said, there was a clear consensus that the climate for women varied from department to department within the School of Engineering. The higher the ratio of female students to male students in particular departments, the fewer gender-related concerns existed. Those participants in Chemical

Engineering felt that they had the fewest difficulties with their male student peers, probably because there was a higher ratio of female students. One participant observed that in Chemical Engineering, the female students in some of the classes were stronger than the male students and that "girls ask and answer questions first." The participants agreed that when and if gender-related issues did arise, they arose primarily with their male peers. It should also be noted that several expressed gratitude for and are currently members of the engineering sorority on campus and claimed that it helped them to deal with being in a predominantly male academic environment.

Participants felt that, in general, they were very fairly and respectfully treated by their primarily male faculty, but they did lament the lack of female faculty who could serve as mentors or models. Some participants pointed out that if and when they have experienced sexual discrimination or harassment it has been when co-oping or job searching. Several expressed concern over the harassment and discrimination they expect they will experience when out in industry. In relation to climate issues, it would be interesting to further explore ways that Catholic universities can develop unique approaches to helping female engineering students deal with sexual discrimination and harassment, and help male students to become more respectful of their female peers.

Andrew Murray, Associate Professor Mechanical and
Aerospace Engineering

As an atheist at the University of Dayton, I struggle with questions of my role and my contribution. I liked the early established notion of this seminar as being "invited to the table." The seminar was meant to both challenge the participants individually and provide a forum for challenging the assumptions of the seminar itself. Nothing made this philosophy more clear to me than the first day we met. I announced my statement of belief. "I am an atheist." This comment passed relatively uneventfully. Then I announced my professional interest. I am a kinematician. "What is a kinematician?" Father Heft asked. "Kinematician," I offered, "comes from the word kinematics. It is the study of motion without regard for the forces that cause it." Father Heft then re-peated this back to the group, "The study of motion without regard for the forces that cause it." Then he added: "How like atheism!"

This challenge, to me, was an important aspect of the seminar. The statements of the Marianist approach to education, at least in the form that I had been exposed to them, made me uneasy. They ask faculty to educate for formation in faith (why is this one always listed first!?), provide an integral quality education, educate in the family spirit, edu-cate for service, justice and peace, and educate for adaptation and change. By the time seminar was over, I perceived these statements about education as a challenge to us faculty in the School of Engineer-ing. The question "How do we fit into this approach?" simply had not occurred to me. And, ultimately, the idea that we can reflect on this question and say, "Here is how we contribute and here is the part of the mission we cannot deliver on because it is not appropriate for a School of Engineering to do so"—that we can say this is where we want to be. I appreciate the way this seminar forced me to start rethinking that relationship.

The last time I participated in a "class" that included literature, history, music, and so forth, was 1989, the year I finished my B.S., more than fifteen years ago. Well, here I was with weekly readings where, not only did I actually have to do the reading, but I had to reflect on it and be prepared to discuss it. I was back in the humanities classroom! To me, this made the humanities/engineering division very clear. Our classrooms are just different places. Also, these humanities people were truly interested in engineering and in helping to produce good engi-neers. But the way they spoke and the problems they identified as im-portant were different than those we engineers would identify. Our language was different. Our attitudes were different. Our lessons were different. These observations comprised the genesis of my research.

My research focuses on the learning and thinking to develop a cross-disciplinary course with Brad Kallenberg, a Christian theologian, titled By Design. Framing the course are two guiding statements.

1.   There is no single, right answer in ethics. There are entirely wrong answers. But, within the range of roughly acceptable responses, each proposed answer must be evaluated for its relative satisfactoriness.

2. There is no single, correct design. There are entirely wrong designs. But, within the range of roughly acceptable solutions to a design problem, each proposed solution must be evaluated for its relative satisfactoriness.

Here is the basic idea for the course. We mix a classroom full of engineering and humanities students. We engage them in the design process repeatedly and in a variety of contexts, most importantly aimed at having them explore "bigger pictures" associated with technical design. Here's an example of an in-class project: we show the students a picture of a river and ask them to design a transport system. There is, of course, a large body of technical knowledge many of them are missing to complete this. But are questions about the technical the only questions that should arise? What about questions such as these: Where is this river located? (Perhaps a border between warring nations.) What is being transported? Why would anyone want to get across? Who owns the land on either side? (Public strips next to waterways are a recent idea.) Who is paying? Is money a metric? Who gets to cross for free? What year is it? How pretty should it be? What is served by it being beautiful?

Here's another project example, which is a little more involved: Develop a map that captures the character of, say, West Dayton (a poor neighborhood) and West Oakwood (a wealthy neighborhood). Can the map show access to some set of resources? Household wealth? What other metrics can we use? How do we portray this on a map?

We are working on a good example where the students need to design a series of monuments, each posing different challenges. For example, the first monument would be to their favorite TV show or movie, something with which they are intimately familiar. The second monument would be to something they can learn about, say a specific person or event that they probably haven't covered in a previous course. The third monument is to something more abstract, say, thermodynamics. Given the abstract character of this third monument, we hoped that the students would simply soak up information and eventually pull some of their imaginative ideas together.

We're thinking the final project will be this: Design a chapel to go in a specific classroom in Kettering Labs (our engineering building). The students are not allowed to change the floor plan, but will have many freedoms beyond that. Each team will, hopefully, have a Marianist brother as a client with whom to discuss the philosophy and bounce ideas around about embodying them in a physical space. Of course, they also have plenty of examples around campus too.

"But Drew," someone might well ask me, "you're not a bridge designer, you're not a monument or a chapel designer. How will you grade this?" From teaching design in a more technical setting, I know to look for evidence of the process being engaged, probably through journals or portfolios—but we'll see. This type of evaluation would constitute a large part of the grade. The rest of the grade comes from the panel—clients, if you will. The students then "justify" their designs to the panel. For example, on the chapel project, the students will justify their choices in capturing philosophies, aesthetics, and so on, to a panel of Marianists.

The goal is to produce in students the desire to ask questions like: In what infrastructure is this problem situated? Who are the constituents and affected parties? Who benefits from the design and who should benefit? Who is profiting and who got screwed?

How do we not benefit from having engineers that: tune their ears to outside voices; rethink the metrics they might use to evaluate design; realize the affected parties aren't just the clients or purchasers; realize they both alter and are altered by political, ecological, economic, and social systems; and, ultimately, view engineering as a society-transforming practice? Forgive our ambitiousness on this one. I appreciate that this is a three credit course we have yet to offer.

In summary: The fact that a Christian theologian and an atheist engineer share a common vision and are pursuing it, well . . . that, I think, is something beyond the understanding of even a kinematician.

APPENDIX A
SYLLABUS FOR ENGINEERING SEMINAR

Winter Semester 2005
Monday 3–5 PM

January 24: Meeting with Mark Schwehn, *Exiles from Eden: Religion and the Academic Vocation in America* (Oxford, 1993).

January 31: History and Current Situation of Catholic Higher Education in the US:
Reading of "Catholic Institutions and Catholic Identity," chap. 4 of Peter Steinfels, *A People Adrift: The Crisis of the Roman Catholic Church in America* (Simon & Schuster, 2003). The author, Peter Steinfels, will be present for the discussion (this semester he is Distinguished Visiting Professor for UD's Ph.D. program in American Catholicism).

February 7: History of Engineering and Engineering Education:
Machine-Age Ideology (chap. 2, "Engineers and Efficiency," and chap. 8, "Roads Not Taken"), by John M. Jordan (University of North Carolina Press, 1994). An excerpt from Brad Kallenberg's forthcoming book on engineering ethics dealing with engineering education in nineteenth-century Germany; L. E. Grinter, "Interim Report of the Committee on Evaluation of Engineering Education," *Journal of Engineering Education* 45 (September 1955): 25–60 ; E. S. Taylor (Chair of the MIT Committee in Engineering Design), "Report on Engineering Design," in *Journal of Engineering Education* 51, no. 8 (1961): 645–60.

February 14: Further Materials on the History of Engineering:
"Episodes in the History of the American Engineering Profession," by Bruce Sinclair, in *The Professions in American History,* ed. Nathan Hatch (University of Notre Dame Press, 1988); Edwin Layton, "Mirror-Image: The Communities of Science and Technology in

Nineteenth Century America," in *The Engineer in America* (University of Chicago Press, 1991). John Heitmann, UD professor of history will be with us and will provide an additional reading. He teaches a course in the history of civil engineering.

February 21: Gender and Engineering:
Harriet Zuckerman, "The Careers of Men and Women Scientists: A Review of Current Research," in *The Outer Circle: Women in the Scientific Community,* ed. Harriet Zuckerman, Jonathan R. Cole, and John T. Bruer (W. W. Norton, 1991); Sherry Turkle, "Computational Reticence: Why Women Fear the Intimate Machine," in *Sex/Machine,* ed. Patrick D. Hopkins (Indiana University Press, 1998); Walter Ong, S.J., *Fighting For Life: Contest, Sexuality and Consciousness* (Cornell University Press, 1981), chap. 4, "Academic and Intellectual Arenas."

February 28: A Catholic Vision of the Intellectual and Professional Life:
"Characteristics of Marianist Universities" (1999), available at http://campus.udayton.edu/~amu-usa/pdfs/characteristics.pdf; Lawrence Cunningham and Mark Roche, *The Intellectual Appeal of Catholicism and the Idea of a Catholic University* (University of Notre Dame Press, 2003).

March 7: Engineering, Society, and Design for Humans:
Thomas Hughes, *Human-Built World: How to Think about Technology and Culture* (University of Chicago Press, 2004), specific pages to be determined. The Ideo Organization, "The Power of Design," *Business Week,* May 17, 2004, and "Design Gets Real," *Newsweek,* October 27, 2003, plus *Nightline* "The Deep Dive," orig. air date July 13, 1999.

March 14: Engineering as a Vocation:
A chapter from Brad Kallenberg's forthcoming book on engineering ethics.
Several short readings on the notion of vocation and profession

March 21: Spring/Easter Break.

March 28: Spring/Easter Break.

April 4: Engineering and Problem Solving:
Discussion of Bill Vaughn Koen, *The Method: Conducting the Engineer's Approach to Problem Solving* (Oxford University Press, 2003), pp. 1–110.

April 11: Further readings on the Nature and Mission of a Catholic University:
"Characteristics of a Marianist University"; Mark Roche, *Intellectual Appeal* (see Feb. 28 above). The Three Column Exercise: Catholic Theological and Marianist themes; Education Ramifications; Applications to the School of Engineering.

April 18: Discussion of Department of Defense Funding and Research at a Catholic University:
James Heft, "The American Bishops' Pastoral on War and Peace," in *The Catechist* (Oct. 1983); "Report of the Review of Research Committee at the University of Dayton," September 27, 1985; *UD and Star Wars,* a collection of essays, March 1986 (published at UD).

April 25: Second round of discussion concerning the Three Columns, with particular focus on the third column. Further discussion on research projects.

May 2: Last Monday meeting of the seminar.

Given the expectation for working over the summer on a publishable article, course innovation, or curricular changes, it could be useful to think about meeting two or three times over the summer to share ideas about our research projects, how they are going, what problems we might be encountering and just provide mutual encouragement and support.

APPENDIX B
THE THREE COLUMN EXERCISE

Traditional Catholic Language	Current Reflections on the Educational Mission of the University of Dayton	Ramifications for the School of Engineering
1. Faith and reason	1. Not only empirical and logical, but also intuitive, committed, and love-based knowing; also the relationship between freedom and truth.	1.
2. Dignity of each person (created in the image and likeness of God)	2. Individual and society: social justice advocacy for the poor and voiceless	2.
3. Unity of all knowledge (both creation and redemption are from God)	3. Beyond "perspectivalism" to search for wisdom through interdisciplinary approaches	3.
4. Both scripture and tradition	4. Centrality of the Word (*logos*): primacy of human participation and interpretation; centrality of history (continuity) and creativity (adaptation)	4.
5. Sacramentality	5. Valuing the tangible and the particular; issues of creation, ecology, and bodiliness	5.

Traditional Catholic Language	Current Reflections on the Educational Mission of the University of Dayton	Ramifications for the School of Engineering
6. Society of Mary themes:	6. Educational philosophy:	6.
a) holy family	a) Education in and through community	a)
b) brothers and priests together	b) Collaboration; non-elitist. liberal and professional education	b)
c) Marian dimension	c) More capacious form of reasoning; perceptive; embodiment; capacity for relationships	c)
d) forming apostles	d) Transformative agents; leadership	d)

APPENDIX C
SUMMARIES OF ADDITIONAL SUMMER RESEARCH PROJECTS

Engineering Courses for Non-engineers

The role of a Catholic university is to properly blend the discovery of knowledge by students, the transmission of knowledge to students, and the character and religious development of students. For years, engineers have benefited from the general education component of their formal education. Non-engineering students could benefit personally and professionally from an engineering component to their formal education. The focus of this research was to provide a summary of attempts at U.S. colleges and universities to offer engineering courses to non-engineers. The categories include everything from single courses that count for general education credit to graduate teacher education programs (certificate and degree). A brief survey of literature on engineering courses for non-engineering students is presented. The topic is introduced and a history of relevant publications is given. Characteristics of previous courses and programs are presented.

Robert J. Wilkens

New Course: Life and Technology

My project centered on the research for and development of a new course, HST 348 Life and Technology, which included broad themes and particular topics of direct relevance to the seminar. The main objective of the course is to examine how life and technology have been mutually understood rather than seeing them as opposites. Topics include life and mechanical philosophy; energy, work, and life (industrialization, thermodynamics and the mastery of nature); cybernetics; biotechnology and genetic engineering; bioinformatics; and automata and robots. Technology has been celebrated as the source of human liberation and derided as the creature that may one day destroy us but people often overlook that their conception of themselves and of life itself may have been informed by or even dependent upon technology. The universe has been viewed as an organism and as a great clock. Genes are considered as lines of code or information for a view of bodies that can be molded surgically and "engineered" to be improved. We fear "playing God" through biotechnology. The seminar deepened my understanding of how engineers contributed to the built environment in which we all live. Engineers help build a way of life rather than merely creating a power grid or developing other systems. It was also important for enhancing my appreciation of the importance of religious values and traditions regarding technology and life—from the metaphor and model of God the designer and universal clock-maker to the sanctity of life and the sense that things have a nature that could be destroyed by engineered manipulation. Life and Technology examines some broad themes but nearly all of them were addressed in the discussions of the seminar and the course was much better for it.

Brad Hume

NOTES

1. The first three seminars were entitled Ethics, Religion and the Professions; The Social Sciences and Catholicism; and Science and Religion. Descriptions of the first two seminars have been written and published: (1) "Ethics and Religion in Professional Education: An Interdisciplinary Seminar," *Current Issues in Catholic Higher Education* 18, no. 2 (Spring 1998): 21–50; also included in *Enhancing Religious Identity: Best Practices from Catholic Campuses,* ed. Irene King and John Wilcox, 175–99 (Washington DC: Georgetown University Press, 2000). (2) "A Study of Catholicism: An Interdisciplinary Faculty Seminar," *Horizons* 29, no. 1 (Spring 2002): 94–113.

2. The Marianists, Dayton's founding religious congregation, are a Catholic religious order founded in France at the beginning of the nineteenth century.

3. I have given drafts of this presentation to all the members of the seminar. Some members may recall other moments as more significant than the ones I am going to describe, but all thought the conversations I decribed were important.

4. Quoted in John M. Jordan, "Engineers and Efficiency," chap. 2 of *Machine-Age Ideology* (Chapel Hill: University of North Carolina Press, 1994), 61.

5. Jordan, "Implementation and Redefinition," chap. 8 of *Machine-Age Ideology,* 201.

6. The Ideo design process is similar to many others: identify the need, observe and learn, brainstorm and visualize, design and build, test, and then repeat. What is different about this process is that the originators of it have learned how to engage people in a design process that is both fun and capable of inspiring a significant amount of learning about their designs. One of the founders of the process, Tom Kelley, recommends that engineers "fail often to succeed sooner."

7. Summers's speech to a conference on women and science, hosted by the National Bureau of Economic Research had been available at http://www.president.harvard.edu/speeches/summers_2005/nber.php, but has been removed from the website. Subsequent statements can be found at http://www.president.harvard.edu/speeches/summers.php.

Chapter Two

The Theological Origins of Engineering

BRAD J. KALLENBERG

Knowledge of our roots can sometimes help us figure out how we *ought* to proceed. Many claim that engineering began in ancient antiquity with the Egyptian pyramids, Archimedes' inventions, or the Roman aqueducts. Others give contemporary engineering a more recent history, tracing its origins to the Industrial Revolution or the Enlightenment. Yet what is often overlooked is the fact that contemporary engineering owes part of its identity to medieval monasticism. The advantage of remembering this history is the bearing it has on the questions "What is engineering for?" and "How ought engineering be practiced?"

Michael Davis makes the claim that, in Western thought, engineering has always played second fiddle to science because we in the West have been bewitched by the myth that engineering is nothing but applied science. But engineering is not merely applied science. Engineering has its own distinctive identity. In the first place, Davis claims that engineering can be distinguished from science by the sheer magnitude of the projects undertaken. Constructing a bridge, building a dam, raising a skyscraper are all tasks that require a great deal of cooperation, the sort of cooperation with which the lone inventor or isolated research scientist may have little experience. Of course, if Hobbes is to be believed about the unstable nature of society, the only body capable of the organization and coercive leadership necessary for such large-scale cooperation is the military. Consequently, Davis traces the origins of modern engineering to the seventeenth century, when France boasted an army of 300,000 foot soldiers. For the first time in European history, those foot soldiers who operated the big weapons became organized into a special unit—the *corps du génie*—a term that connotes both the "engines" of war and the "genie" or magic associated with their function.[1]

Although the seventeenth-century corps members are probably best thought of as proto-engineers, the officers of this unit (*officieurs du génie*) began to undergo a formal training (in mathematics, technology, and officer training) whose curriculum differs only slightly from that of today's schools of engineering.[2] Thus what had begun as École des Travaux Public (the School of Public Works), by 1794 became École Polytechnique. This institution is still in operation and its curricula became the model for the first school of engineering in the Americas, namely, the US Military Academy at West Point.

In a moment I will explain why I think Davis's account is incomplete, if not downright wrong. Yet there is much to be said for it. Who could disagree that warcraft was the soil in which even the smallest technological advance blossomed with importance to the end that each was coopted for military use? Famously, the invention of the stirrup secured the superiority of the Frankish cavalry over their more loosely seated opponents in the eighth century just as the development of a better trigger enabled William the Conqueror to utilize crossbows to overpower his Norman opponents in the eleventh.[3]

Moreover, engineering has a distinctive domain of knowledge. As is often the case, with specialization comes a sort of tunnel vision. Perhaps this helps to explain the enduring tendency of engineers to be more enamored with "engineering as an end in itself rather than as a means to satisfying human need."[4] On the one hand, if engineers are descendants of military officers, then they have been trained and bound to do as they are commanded. On the other hand, each generation of engineers is entrusted with a growing and specialized body of knowledge the mastery of which requires successively greater and greater amounts of time and attention. Consequently, it is not surprising that engineers today are easily caricatured as task-oriented folk who are more apt to keep their noses to the grindstone than to trouble with the "why?" questions that seem to lie outside the purvey of engineering itself.

Many (and I among them) feel that the tunnel vision, which is so stereotypical of contemporary engineering, points to a lamentable failing. If "doing one's job" does not automatically indemnify soldiers acting under orders, why should engineers be excused from making ethics or economics or politics their business? While Davis concedes that this tendency may appear troubling, he explains that treating engineering as an end in itself is not identical to tunnel vision and therefore not necessarily a bad thing (though it may be). Although engineering was conceived and birthed by the military, he argues, it matured during the Age of Enlightenment, a time rife with the optimism that scientific learning in every form "would bring peace, prosperity, and continuous improvement."[5] The conclusion Davis wants his readers to draw is that engineers may be somewhat justified in their narrow preoccupation with technical brilliance because the enterprise by its very nature as an offspring of the Enlightenment cannot but serve human need and improve society. Therefore, the concentration with which engineers treat engineering as end in itself in lieu of attention to broader social issues need not tarnish the image of engineering as a morally oriented enterprise.

Davis does have a point. Engineering has an excellent track record in the service of human need. And in large measure, this track record functions as a gyroscope that helps engineering stay on course despite the tangential impetuses of governmental agendas and business "needs." However, I am not convinced that this gyroscope, while absolutely necessary to the continued flourishing of technical expertise, is a sufficient condition for engineering as a whole to maintain its moral bearings. There is a latent ambiguity in engineering's self-understanding. If human life is for increasing market share in a capitalist economy, then designed obsolescence is a reasonable engineering strategy. If human life is for protecting the security of one's people against others, then engineering's four-hundred-year-long allegiance to the military is entirely appropriate. The question, "What is human life for?" has enormous bearing on the practice of engineering.[6] Therefore, engineers may benefit from an account of engineering history that is already steeped in an account of what human life is for.

The story I wish to narrate takes us back to the twelfth-century monastery of St. Victor in Paris. I will argue that what they called "mechanical arts" is a forerunner of what we today call "engineering." Central for my purposes is the *Didascalicon,* written in the 1120s by Hugh of St. Victor. It not only exemplifies a theological model for understanding the identity of engineering, it has bearing on how engineering ought to be practiced.

It is not altogether clear when in history "mechanical arts" becomes recognizable as proto-engineering or when engineers successfully shed their longstanding class stigma. The earlier one looks in ancient history, the more disparaging is the view toward mechanical things. Archimedes (d. 212 BCE) may have saved Athens with his contraptions, but as Plutarch explains, he did so shamefully, fully aware of Plato's "indignation at [mechanical arts], and his invectives against it as the mere corruption and annihilation of the one good geometry."[7] Likewise Plato's contemporary, Xenophon (d. 354 BCE) makes it clear that no true gentleman practiced "mechanical arts." Xenophon reports Socrates' exclamation:

> [N]ot only are the arts which we call mechanical [*banausikai*] generally held in bad repute, but States also have a very low opinion of them,—and with justice. For they are injurious to the bodily health of workmen and overseers, in that they compel them to be seated and indoors, and in some cases also all day before a fire, and when the body grows effeminate, the mind also becomes weaker and weaker. And the mechanical arts, as they are called, will not let men unite with them care for friends and State, so that men engaged in them must ever appear to be both bad friends and poor defenders of their country. And there are States . . . in which not a single citizen is allowed to engage in mechanical arts [*banausikas technas*].[8]

Mechanical arts, in other words, were for slaves.

But what activities fall under the domain of "mechanical arts"? As indicated by Xenophon's words, the mechanics (*banausous*) spent a large part of the day at the foundry (*baunos* was the forge, or furnace). So smithing is implicated as undignified. But evidently warcraft is not. Nor is agriculture. Socrates goes on to assert that these latter activities are for gentlemen.

At first blush, Aristotle's three-fold division of the rational soul into theoretical, practical, and productive reasoning holds more promise for elevating the status of manual crafts. But Aristotle (d. 322 BCE) could not resist falling prey to the hierarchy of the disciplines that gives wisdom (*sophia*) clear priority over intelligence (*phronesis*) and intelligence explicit reign over craftsmanship (*techne*—as in "technology"). To make matters worse, Aristotle also perpetuated the pejorative sense of "mechanical" (*banausous*). So uncontestable is the slur against all things mechanical that in the *Nicomachean Ethics* it is simply translated as "vulgar"![9] Similarly, in the *Politics* he writes that

> any occupation, art, or science, which makes the body or soul or mind of the freeman less fit for the practice or exercise of excellence, is mechanical; wherefore we call those arts mechanical which tend to deform the body, and likewise all paid employments, for they absorb and degrade the mind.[10]

If we leap ahead eight centuries to the close of Plato's Academy—simultaneous with the founding of the Order of the Benedictines in 524 CE—we will discover that among intellectuals not much has changed. Consider Boethius (d. 524 CE), arguably the most significant philosopher-theologian between Augustine of Hippo (d. 430 CE) and Thomas Aquinas (d. 1274 CE). He divides the love of wisdom (*philosophia*) into two disciplines only, theoretical and practical, entirely neglecting to mention productive (mechanical) arts. When Isidore of Seville (d. 636 CE) compiles his encyclopedia a generation later, he acknowledges a number of disciplines that lie outside the classic seven that constitute liberal arts.[11]

Most striking is the inclusion of mineralogy alongside the eminently reputable enterprises of medicine and agriculture. But unfortunately, Isidore did little to improve the social standing of the mechanical arts. His fascination with etymology led him to mistake the Latin *mechanicus* as derived from the Greek *moichos,* meaning "adulterer" rather than from *mechane* (machine) and *mechos* (a means, something expedient, a remedy). To his credit, there is some plausibility for this mistake. Martin of Laon (d. 680 CE) takes Isidore to mean that the ingenuity of a mechanism was akin to the secret doings of an illicit sexual affair:

> from "moechus" we call "mechanical art" any object which is clever and most delicate and which, in its making or operation, is beyond detection, so that beholders find their power stolen from them when they cannot penetrate the ingenuity of the thing.[12]

But of course Isidore's genealogy could not help but accentuate the stigma that afflicted artisans and remind them of their proper place at the bottom of the feeding chain. Perhaps this stigma explains why the Cistercian Order (founded 1098 CE) explicitly forbade "profane" learning and aimed to "make of every monastery a 'school of charity' only."[13] Human life was for the love of God and neighbor, but evidently mechanical arts lay outside the pale of such love.

The situation would change in the twelfth century. Hugh of St. Victor (d. 1142 CE) presents the first cogent challenge to the mechinists' stigma by offering a theological account of the practice of mechanical arts. Granted, Hugh's account was not without rivals: his contemporary, William of Conches (d. 1154 CE), disdained the mechanical arts as merely menial.[14] The difference between Hugh and William lay in their starting points. While William began anthropologically with human knowledge (*scientia*), Hugh began theologically with the doctrine of "sin."

The ancient Greeks explained evil in the world as the residual effect of an eternal battle between the powers of good and evil. Evil was not only conceived as a something, it was an eternal something. Thus, in the beginning was chaos. But Augustine, writing a millennium after Homer and clearly Hugh's hero, could not dignify evil with substance, much less with eternity, for as scripture spelled out, "in the beginning, God . . . ." In other words, in order to affirm monotheism, Augustine was bound to describe evil as having a temporal beginning. And, in order to avoid the conclusion that God created evil, Augustine insisted evil wasn't a substance, but an absence; evil was a defect that entered the picture some time after God had created an entirely good world.

Whence evil? Evil was a distortion in the order of creation effected by a misuse of creaturely freedom. How so? In order for creation to be a uni-verse (rather than a multi-verse), creation embodied a single hierarchy of value.[15] The human soul operates correctly when it ascribes that quality of love appropriate to the object in light of its place on the hierarchy. Augustine (following Plato) considered the order itself every bit as real as the tangible objects that populated the hierarchy of the created world. Evil entered when human beings re-ranked the hierarchy of creation, ascribing an inordinate quantity of love to one or more of the rungs of the hierarchy. In the Apostle Paul's words, "For they exchanged the truth of God for a lie, and worshiped and served the creature rather than the Creator, who is blessed forever."[16] In essence, human mis-valuing was a distortion of the order of creation. The change was very real, although it was a distortion they bore within themselves, for a disordered love is a disordered soul. Thus, disordered human love manifests itself sometimes as greed, other times as jealousy, covetousness, pride, and so on. This condition had the unpleasant consequence of being perpetual, because one could only make moral progress if one possessed a faculty for indexing the progress made. And it was this very faculty, namely love of the Good, that could not be trusted.

But the bad news does not stop with human depravity. Once human beings, viceroys of creation, became incapable of rescuing themselves (*non posse non peccare,* not able not to sin), the creation they were supposed to tend fell under a curse. Christian scripture aptly expresses its undeniable reality:

For the anxious longing of the creation waits eagerly for the revealing of the sons of God. For the creation was subjected to futility . . . in hope that the creation itself also will be set free from its slavery to corruption into the freedom of the glory of the children of God.

For we know that the whole creation groans and suffers the pains of childbirth together until now.[17]

Whether we call this curse "sin" or "entropy" makes little difference for my argument. The fact of the matter is: iron rusts, people sicken and die, and things fall apart.

This then is the theological view of human existence that Hugh inherits from Augustine. As fallen creatures, human beings have forgotten who they are and whose they are. Nevertheless, Hugh writes, "we are restored through instruction, so that we may recognize our nature."[18] God in his redemptive grace and wisdom has intended the very condition of human fallenness as the impetus for human pursuit of Wisdom, a quest which is the "highest curative in life."

> And so arose the pursuit of that Wisdom we are required to seek— a pursuit called "philosophy"—so that knowledge of truth might enlighten our ignorance, so that love of virtue might do away with wicked desire, and so that the quest for necessary conveniences might alleviate our weaknesses. These three pursuits first comprised philosophy. The one which sought truth was called theoretical; the one which furthered virtue men were pleased to call ethics; the one devised to seek conveniences custom called mechanical.[19]

In this passage Hugh asserts that the redemption of the soul is assisted by the practice of "arts" that correspond with all the powers of the soul. Corresponding to the understanding (*intelligentia*) are both the theoretical arts (that is, the contemplation of necessary truths; here Hugh intends theology, physics, and mathematics) and the practical arts (namely, the practice of morality and the cultivation of virtue). Corresponding to knowledge (*scientia*) are all the mechanical arts. These latter have to do with feeding, fortifying the body against harm, and the contrivance of "remedies" for alleviating physical weakness (I.8, p. 55).

Hugh's account is a *"nouveau explicitement,"* a brand new way of thinking.[20] By paying more attention to the doctrine of the human fall into sin, Hugh is able to move beyond his forebears (such as Boethius) and include mechanical arts under God's plan of redemption. Mechanical arts have to do with countering the effects of the curse, just as theoretical and practical arts have to do with countering the effects of human depravity, through the knowing and following of a gracious God on a redemptive path.

Hugh's inclusion of the mechanical arts is no small feat, for "mechanical arts" by his day had evolved into a very broad category. To be specific, mechanical arts was comprised of seven classes of practices: fabric-making, armament, commerce, agriculture, hunting, medicine, and theatrics.[21] These seven name families of practices. For example, "hunting . . . includes all the duties of bakers, butcher, cooks, and tavern keepers," as well as those who actually do the gaming, fowling and fishing (II.25, pp. 77–78). And "armament" included material science, even metallurgy: "To this science belong all such materials as stones, woods, metals, sands, and clays" (II.22, p. 76). With this last move Hugh has managed to embrace even the grimy-faced smithy so consistently maligned for sixteen centuries.

Though fiercely loyal to Augustinian theology, Hugh parts company with Augustine's Platonic division of human arts into physics, ethics, and logic, opting instead for Aristotle's quaternary of theoretical, practical, productive, and logical disciplines. Under Hugh's hand, "productive" arts expands to include all mechanical arts known to him and "logic" alone becomes the special domain of philosophy that governs the consistency within each art and between all the arts.

Hugh argues that, as a theologically legitimate enterprise, mechanical arts were governed by logic every bit as much as were theoretical and practical arts. This means that mechanical arts can be evaluated for how well they aimed at the human Good. Since Hugh could not conceive of any human Good other than that revealed by the divine Wisdom, all of the mechanical arts aim at redemptive love. To cite one example of this redemptive vision at work, Hugh asserts "commerce" as the mechanical art that aims at reconciliation of strangers: "The pursuit of commerce reconciles nations, calms wars, strengthens peace, and commutes the private good of individuals into the common benefit of all" (II.23, p.77).

In Hugh's mind theology and mechanical arts are mutually supportive. The ends of mechanical arts are displayed by the physical things contrived by the artificer. As these ends are theological in nature (they aim at the Good revealed by God), mechanical arts benefit theology by rendering visible invisible things. A bridge is not merely a convenience, it is also a means of cultivating friendship between rival villages on opposing banks. Theology in turn benefits mechanical arts by providing a benchmark for assessing the aptness of its aims. But Hugh is quick to caution against mistaking worldly theology (a theology that moves from human experience to the knowledge of God) for graced theology (a theology that moves from knowledge of God to human experience). In his *Exposition of the Heavenly Hierarchy,* Hugh writes:

> Invisible things can only be made known by visible things, and therefore the whole of theology must use visible demonstrations. But worldly theology adopted the works of creation and the elements of this world that it might make its demonstration in these. . . . And for this reason, namely, because it used a demonstration which revealed little, it lacked ability to bring forth the incomprehensible truth without stain of error. . . . In this were the wise men of this world fools, namely, that proceeding by natural evidences alone and following the elements and appearances of the world, they lacked the lessons of grace.[22]

What are these lessons of grace? For Hugh grace is not something added on top of nature, but something that permeates the world and with which human beings may keep step. "Grace," writes Hugh, is the powerful medicine perpetually offered by God "to illuminate the blind and to cure the weak; to illuminate ignorance, to cool concupiscence; to illuminate unto knowledge of truth; to inflame unto love of virtue."[23] In contrast, worldly theology is like tugging at one's bootstraps. It reveals little, and therefore has little to say to mechanical arts, precisely because it ignores God at the outset. Worldly theology begins with an empirical study of "pure nature" and then attempts to reason up toward the possible existence of a divine realm. But graced theology unblinkingly assumes that creation is already shot through with the presence of God. Wherever one points is God's world. Human beings live as creatures under a creator whose divine wisdom is the archetypal exemplar of creation.[24] Granted, evil happens. But the undeniable fact of evil only serves to corroborate strongly the biblical story that human beings are fallen creatures inhabiting a cursed world. The Fall and its effects are universal in scope (how could it be otherwise?). Nevertheless, even in their fallenness, human beings are redeemable in the pursuit of divine wisdom by means of exercising theoretical, practical, and mechanical arts. The final end of mechanical arts is reunion with God through the pursuit of divine wisdom as well as the alleviation of physical weakness stemming from the cursedness of the created world.

Three conclusions can be drawn from Hugh's *Didascalicon*. First, in Hugh's day the growth of technology was already noticeable enough to require a fresh classification long before the seventeenth century that figures so prominently in Davis's story of the identity of modern engineering.[25] Second, Hugh's account shows that despite our penchant for separating engineering and theology, a theological account of the mechanical arts was possible. This is not to say his account is more persuasive than Davis's, only that engineering need not be excluded from a theological account of human life.

The third conclusion to draw from the *Didascalion* is that, for a careful thinker such as Hugh, a theological account was the only account that was broad enough to encompass all he had learned from Plato (especially the *Timaeus*), Aristotle, Augustine, Boethius, Varro, Quintilian, Isidore, and others. His strategy was to absorb all the pertinent sources into a master theological narrative. We moderns tend to be suspicious of such a methodology. We are more accustomed to reconciling diverse views (if they can be reconciled at all—and it has become increasingly in vogue to assume an incommensurable plurality of views) by reducing all the views to their greatest common denominator. Of course such a reductive methodology means that those tenets distinctive to specific religions such as Christianity or Islam or Judaism must be surrendered in the name of peaceful coexistence with its rivals. But then what is left? The greatest common denominator, it would be argued five centuries later, was the notion of "pure nature." However, this notion was simply not available to Hugh for two reasons.

In the first place, as Henri de Lubac has convincingly argued, the perspective of present-day historiographers may be blurred by three hundred years of (Cartesian) dualism that wrongheadedly presupposes it even makes sense to speak of "pure nature." In this thoroughly secularized vision, notions of "grace," "spirit," "calling," and the "supernatural"—if they have substantive content at all—are concluded to be mere add-ons to a presumably more basic concept: "pure nature." But this could not have been the Christian outlook in its previous fifteen centuries. The creation of human beings in God's image had sweeping ramifications for understanding for what human life was intended. Medieval Christians simply took it for granted that human beings were "destined to live eternally in God, to enter into the inner movement of the Trinitarian life and to bring all creation with [them]."[26] In other words, for medieval believers, "nature was made for the supernatural" and cannot even be conceived, much less explained, without it.[27] The inseparability of natural and supernatural typified the medieval Christianity and is given its most eloquent expression by Augustine who included all creation in the "us" of the famous opening to his *Confessions*: "You have made us for yourself, and our heart is restless until it rests in you."

In the second place, until late in the seventeenth century, "natural philosophy" covered much of the same domain as "Christian theology." It is a colossal misunderstanding to think natural philosophy studied "pure nature" while Christian theology studied a putative "supernatural" realm. On the contrary, both disciplines were overlapping responses to the created world (and it was seen as creation, rather than something else).[28] Historian of science Margaret Osler writes,

> Medieval natural philosophy was conditioned by theological presuppositions, and its conclusions pertained to important theological issues. Discussions of the causes of things, for example, included questions about the cause of the world and revolved around the issues of the divine creation of the world. Discussions of matter and change had implications for the interpretation of the Eucharist. Discussions of the nature of animals and how they differ from humans had direct bearing on questions about the immortality of the human soul.[29]

Osler's words point to the fact that medievals could not separate efficient and material causes from final causes in their explanations. Christian theology and natural philosophy had overlapping domains because in a created world both disciplines had the same final cause (namely, union with the creator God). It is only after the Enlightenment project abandons Aristotle that subsequent moderns are tempted to read mere instrumentalism—efficient causes taken in isolation from final causes—back into medieval thought.[30]

Perhaps an example can make this clearer. The requirement that monks devote themselves to work is widely acknowledged. What is contested today is whether the Benedictine motto, "work is prayer" (*laborare est orare*), originally reflected a sacramental rather than an instrumental view of work. For his part, historian Jacques Le Goff maintains that monks worked hard, intentionally trying to improve their efficiency with machinery (such as the water-powered mill constructed at Saint Ursus at Loches in the sixth century) so as to free up time for the essential thing: *opus Dei*, namely contemplative prayer.[31] In other words, Le Goff can see water mills for saving time but not for worship. Does Le Goff get history wrong?

Le Goff's history is not so much wrong as it is monochromatic. Did water mills save time? Of course. But where Le Goff sees in black and white, Hugh sees in resplendent color. For Hugh, mechanical arts yielded artifacts (and processes) that were inherently sacramental because they rendered visible the end of mechanical reasoning, which in its exercise was simultaneously natural (namely, the alleviation of physical weakness) and supernatural (namely, the journeying toward reunion with divine wisdom).

The strength of Hugh's theological account is that it supplies what nontheological (what Hugh called "worldly") accounts could not as easily do, namely, thick description of the final end toward which all human activity aims. In sum, it was by the "lessons of grace" that Hugh was able to see the physical world under both the aspect of the supernatural and the aspect of the natural. Accordingly he described mechanical arts as guided by a dual end. The supervening supernatural end is this: the exercise of mechanical reasoning is part of the journey toward reunion with God. The subvening natural end is this: mechanical artifacts are for the alleviation of physical weakness that is the consequence of living in a fallen world.

NOTES

Many thanks to Terry Tilley, Therese Lysaught, Aaron James, John Heitmann, and Joe Jacobs for their many helpful comments on an earlier draft of this essay.

1. Michael Davis, *Thinking Like an Engineer* (New York: Oxford University Press, 1998), 10.

2. Ibid.

3. Ton Meijknecht and Hans van Drongelen, "How Is the Spirituality of Engineering Taught or Conveyed?" (paper presented at the Designing Engineering Education: Mudd Design Workshop IV, Claremont, California, July 10–12, 2003).

4. Davis, *Thinking Like an Engineer,* 10.

5. Ibid., 15.

6. Alasdair MacIntyre, *After Virtue: A Study in Moral Theory,* 2nd ed. (Notre Dame, IN: University of Notre Dame Press, 1984).

7. Plutarch, "Marcellus," in *The Lives of the Noble Grecians and Romans* (New York: The Modern Library, 1932), 376.

8. Xenophon, "The Economist of Xenophon," trans. Alexander D. O. Wedderburn and W. Gershom Collingwood, *Bibliotheca Pastorum,* vol. 1 (New York: Burt Franklin, 1971), IV.2, pp. 22–23.

9. Aristotle, *Nicomachean Ethics* 4.2, 1123a, available from the Perseus Digital Library, http://www.perseus.tufts.edu. See also *Eudemian Ethics* 1.4, 1215a, and *Rhetoric* 1.9, 1367a.

10. Aristotle, "Politics," in *The Complete Works of Aristotle,* Bollingen Series, ed. Jonathon Barnes (Princeton, NJ: Princeton University Press, 1984), 8.2, 1337b.

11. The classical quadrivium was comprised of arithmetic, music, geometry, and astronomy, while the trivium was comprised of grammar, rhetoric, and logic.

12. Hugh of St. Victor, *Didascalicon,* trans. and ed. J. Taylor (New York: Columbia University Press, 1961), p. 191, n. 64.

13. Arthur O. Lovejoy, *The Great Chain of Being: A Study of the History on an Idea* (New York: Harper, 1960), 4.

14. Hugh of St. Victor, *Didascalicon,* from the introduction, p. 4.

15. This view was nothing new; the ancient Greeks pointed out that there must be some reason why it is easier to step on a cockroach than to put down a horse, namely, that horses have more inherent worth—are higher on the hierarchy of being—than roaches.

16. Romans 1:25.

17. Romans 8:19–22.

18. Hugh of St. Victor, *Didascalicon* I.1, p. 47.

19. From Hugh's *Epitome Dindimi in philosophiam,* cited in Hugh of St. Victor, *Didascalicon,* 12.

20. Marie-Dominique Chenu, "Arts «Méchaniques» Et Œuvres Serviles," *Revue des sciences philosophiques et théologiques* 29 (1940): 314.

21. "Theatrics" may seem like a stretch to include under mechanical arts, but Hugh made the list seven in number so that it matched the perfection of the seven liberal arts. Besides, under theatrics Hugh envisioned any coordinated activity of a group of people. Not just drama, but marching bands and gymnastics would fit under this heading. Had Hugh lived to see Ford's assembly line, he surely would have treated it as a type of theatrics.

22. Cited in Hugh of St. Victor, *Didascalicon,* from the introduction, p. 35.

23. Quoted in Aage Rydstrom-Poulsen, *The Gracious God: Gratia in Augustine and the Twelfth Century* (Copenhagen: Akademisk Forlag, 2002), 206.

24. Hugh of St. Victor, *Didascalicon,* from the introduction, p. 13.

25. Ironically, Hugh does not mention watermills in the list of technologies he describes, though Paris is not far from St. Ursus monastery at Loches in the Loire Valley, the site of a sixth-century mill. This may only signify that the Parisian Hugh was a city boy who had no firsthand experience with mills.

26. Henri De Lubac, "Internal Causes of the Weakening and Disappearance of the Sense of the Sacred," in *Theology in History* (San Francisco: Ignatius Press, 1996), 230.

27. Ibid., 231.

28. Margaret J. Osler, "Mixing Metaphors: Science and Religion or Natural Philosophy and Theology in Early Modern Europe," *History of Science* 35 (1997): 91.

29. Ibid., 92.

30. MacIntyre, *After Virtue.*

31. Jacques Le Goff, *Time, Work, and Culture in the Middle Ages,* trans. Arthur Goldhammer (Chicago: University of Chicago Press, 1980), 80.

PART 2

Building the Bridge

Chapter Three

A Catholic and Marianist Engineering Education

KEVIN HALLINAN AND

MARGARET PINNELL

The School of Engineering at the University of Dayton (UD), a Catholic and Marianist University, boasts large enrollments of 1,300 undergraduate and 350 graduate students out of a total of 7,000 undergraduates and 3,000 graduate students. It also boasts a faculty very active in research, which, under the umbrella of the University of Dayton Research Institute, is funded at a level of $100 million per year. In our region, we are looked at as one of the premier engineering programs in terms of the quality of the graduates we produce.

In the last decade, the University of Dayton has sought to better articulate the impact of its Catholic and Marianist traditions, and faculty have been challenged to embody these traditions. University mission statements and unit strategic plans have also evolved to make better connections. In this context, our paper explores the historical and present connections to these traditions, and then more importantly presents a vision for better integration of them into the education of our students. The visioning really represents an early foray into thinking about greater embodiment of mission into the engineering education at Catholic universities. Finally, we envision what a specific application of the principles to a course in thermodynamics would look like and consider extension to all engineering courses.

WHAT HAS CATHOLIC AND MARIANIST MEANT IN THE PAST?

Engineering found its foothold at the University of Dayton in the early 1900s, with the establishment of the Departments of Chemical and Electrical Engineering in 1911 and Mechanical and Civil Engineering in 1914. This addition was relatively early for Catholic universities, owing to the desire of the Marianists to provide "practical" education to the Dayton region, which was already well on its way to becoming a vibrant technology center. But, in its inception, there was a bifurcation of Catholic identity and the technical training provided to engineers. For example, in the 1914 University of Dayton catalog, twenty-two periods of class (forty-five minutes each) per week were required. These included a 1½ hour course per week in Christian Doctrine that was taught only by vowed religious. In a four-year sequence, this course was sequentially focused on Catholic morals, Catholic dogma, apologetics, and rational theology. The only other non-engineering courses required were mathematics, sciences, English, German, and French. According to our most senior alumni, Catholic was the domain of the theology education our students received.

This curricular scenario changed little through the 1950s. To be fair, however, the "helping" environment created by the Marianist educators did influence the teaching of the lay faculty to create a student-centered and nurturing educational environment. In the 1960s, the University of Dayton, like other Catholic universities, sought to flex its intellectual muscles relative to secular universities. Thus, there was greater pressure to achieve success in engineering relative to the same metrics of success employed by secular universities.[1] The formation of the University of Dayton Research Institute, fundamentally connected to the School of Engineering, in response to Wright-Patterson Air Force Base's desire to contract out research services, is illustrative of the university's response to this pressure. Further, the university established a Ph.D. program in 1973 because the Ph.D. was said to represent "the highest recognition of scholarly achievement."[2]

Vatican II may have also led to substantive changes that affected the education of Dayton's engineering students. The introductory section of the 1973–74 Bulletin rarely spoke of Catholic and Marianist traditions, and instead emphasized "varied religious, social, and cultural opportunities," and described UD as a "church-related institution of higher learning, where academic freedom was sought." Such language conveys a strong sense that the university was not sufficiently attending to its Catholic and Marianist heritage.

The University of Dayton apparently was not unique. As described by Phillip Gleason, during this time Catholic universities appeared to have lost sight of their identity.[3] At the same time, the number of vowed religious experienced a dramatic decline. Increasingly, church organization and university administration shifted to lay leadership. Yet in the initial years after Vatican II, short of the continuing caring and nurturing environment created by the faculty in engineering, Catholic and Marianist mission continued to remain the responsibility of the dwindling numbers of vowed religious.[4]

In the 1980s, the educational "heart" of the University of Dayton was defined to be the humanities, and a new, more integrated general education program was established. In its development, as described through conversations with its developers, only a slight Catholic and Marianist filter was applied, seen primarily in the creation of a Humanities Base curriculum, for which the organizing question was "What does it mean to be human?" Still in existence, the curriculum more explicitly integrates Catholic and Marianist themes.

As documented by David O'Brien, the 1980s and '90s saw a fervent attempt by Catholic universities to define themselves more clearly.[5] The University of Dayton's articulation of this rejuvenation was its Vision 2005 document in the mid-1990s, which was described as a roadmap for transforming all elements of the University of Dayton through a pattern of lay ownership of mission. Faculty in the professional schools were then challenged to think about how they might connect to this mission. But it did not immediately happen. For example, the

1997 School of Engineering response to the Vision 2005 document envisioned primarily an improved scholarly reputation relative to secular universities, primarily calling for more resources in support of graduate students. This request was dismissed as being unrealistic by then President Raymond Fitz, S.M.

Since 1997, however, there have been profound changes. Our students' and recent alumni understanding and commitment to ethical behavior in a national survey instrument used for assessment is number one in the US. Further, after our ABET (Accreditation Board of Engineering and Technology) accreditation visit in the fall of 2004, the ABET team lead who was finishing twenty-five years of service with ABET, in closing with our dean's staff, said with tears in his eyes that he and his team "had been blessed by the warmth of our students, faculty, and staff, and by the deep commitment to educating our students." The mechanical engineering program evaluator, with whom we primarily interacted, a committed Catholic at a small state college, likewise offered, "It was nice to see a university with stated values that are lived." Within this context, this chapter discusses the ways in which engineering education at the University of Dayton has evolved and might continue to evolve to embrace its Catholic and Marianist traditions.

WHAT DOES CATHOLIC AND MARIANIST MEAN NOW?

Until the relevance to the curriculum of being Catholic and Marianist could be established, it was impossible to envision how the engineering curriculum might be influenced by this mission. Through the early 1990s, this was the case. While one could honestly say that there was something different here, that difference was hard to describe even for some of the vowed Marianists. In 1996, significant effort was invested into the creation of a document describing the characteristics of Marianist Universities.[6] The resulting publication was a collaborative effort between the three Marianist universities in the United States (University of Dayton, Chaminade University, and St. Mary's University). Five characteristic elements of the Marianist approach to education were identified, including:

- educate for formation in faith
- provide an integral quality education
- education in family spirit
- educate for service, justice, and peace
- educate for adaptation and change

The implicit impact of these on education was detailed in the document. Education for formation in faith means that the Catholic character of the university is vital and helps to produce distinctive graduates. It also means that both faith and reason are emphasized and that faith must be in dialogue with and in service to culture. By an integral quality education, education of the whole person is implicit, as is the search for knowledge. Education is also seen as not simply the domain of the classroom, but of the whole of the experience. Education in family spirit starts with a climate of acceptance, and a recognition of the importance of building community both within and outside of the university. Education for service, justice, and peace stresses an almost vocational perspective for work and particular concern for the poor and marginalized. These elements are not seen as separate from the curriculum. Finally, relative to education for adaptation and change, there is a call for education that envisions changes to culture and adapts accordingly, prepares graduates to live in a pluralistic society, and develops critical thinking skills in the search for truth.

As far as what is meant by Catholic, there is much variation of opinion. What is consistent among all Catholic institutions is service to the community—local, regional, and global.[7] The constitution of the communities served and the nature of the learning from such experiences is of course varied. In terms of education and scholarship, there is also wide variation. For example, Franciscan University of Steubenville, Ohio requires a strict adherence to a distinctively conservative Catholic teaching in all areas of college life. However, this strict adherence to a certain form of religiously affiliated education may arguably stifle the development of critical thinking. Further, such a model may not fare well using success metrics employed by secular universities.

Others have suggested broader interpretations of a Catholic university that permit distinctiveness relative to secular institutions, while at the same time offering a competitive product relative to academic/ professional as well as public/government organizations.[8] For engineers graduating from Catholic universities, "a competitive product" means simply that, beyond preparation in accordance with the mission of the institution, they must be both employable and prepared for advanced study within and outside of their profession. Catholic universities that have striven for rich scholarship have argued that the search for knowledge in itself is good, as all knowledge is a revelation of God.[9] Such a pedagogy fosters acceptance of disciplinary separation and national success according to metrics used by secular institutions that reward demonstration of disciplinary expertise. Others have suggested the establishment of Catholic studies or Catholic intellectual tradition studies for all students,[10] going beyond the typically nonintellectually based campus ministry activities. James Heft, S.M., has further advocated for incorporation of the Catholic intellectual tradition into the research and teaching of all disciplines.[11] Along these same lines, historian James Turner suggests that "Catholic colleges have seldom encouraged their students to think seriously and flexibly about their faith."[12] David O'Brien offers that Catholic education should educate to create *disciples and citizens*.[13] Graduates would then be filled with a sense of religious inspiration to serve the world positively. Finally, Patrick Byrne has posited that a vocational exploration should be part and parcel of the education of students (and faculty) in all disciplines.[14]

A Catholic education encourages commitment to a few basic principles. First, it emphasizes a reliance of faith and reason. This implies that knowing is not simply logical and intuitive. Thomas Groome has suggested that this commitment also implies that a humanizing education is an "aspect of the work of our salvation."[15] It can also have a love-based and intuitive component. For engineers, this linkage between faith and reason would translate to "passion" behind the profession, to the *why* for being an engineer, encouraging an education that inspires students to view their profession as a service to the common good.

A Catholic education also recognizes the dignity of each person. For engineers, such an education would require consideration of the effects their labor has on the whole of society and would implicitly define the relationships they ought to have with colleagues and clients. As Groome suggests, Catholicity is an inclusive concern.[16]

Third, a Catholic education recognizes the sacramentality of creation. For engineers, this facet of education inevitably leads to a commitment toward protecting the gifts of creation. Sustainability, as an integral philosophy for engineering, seems a likely end-goal of this vision of sacramentality. As Thomas Landy describes, sustainability, since it focuses primarily on human, social goods, stresses a "responsibility and a greater sense of common good, and relies on stewardship more than on boundaries between humans and nature."[17]

Fourth, a Catholic education recognizes the unity of knowledge. From a professional school standpoint, this would place importance both on establishing an environment that provides students with breadth of knowledge and on connecting such knowledge.

Finally, Catholic education considers that both scripture and tradition are important. While scripture may reveal fundamental underlying truths, tradition recognizes their meaning in different periods of history. Relative to engineering, faculty generally have little understanding of both the history of the specialized knowledge they teach and the history of engineering in practice. Engineering for the world involves continuous *feedback loops*. We design, we make, we learn, we redesign. We design, we make, we change society, and we redesign. Rarely does engineering education provide even a glimpse of these feedback loops or a glimpse of the amazing impacts—not all of which are positive—of our efforts to either engineering or non-engineering students.

Collectively then, as we see it, the theological educational impacts of our Catholic and Marianist traditions applied to engineering would at least include the following elements:

- developing and living the passion for the profession (the faith behind the reason)
- integration of knowledge and problem solving across disciplinary boundaries (unity of knowledge)
- building and educating for community (educate in the family spirit)

- working as servant leaders striving for justice within and outside of profession
- a focus on sustainable engineering practices (sacramentality of creation)
- preparing for adaptation and change

It is interesting to note that none of these educational outcomes is included in outcomes required by ABET. While ABET includes outcomes related to multidisciplinary teamwork, continuous learning, knowledge of the social and cultural environment in which engineers work, and ethical understanding, such outcomes are pale reflections of those described above.

THE GOALS OF A CATHOLIC ENGINEERING EDUCATION

Developing and Living a Passion for the Profession of Engineering

As the Catholic and Marianist voice has gained strength, there have been numerous faculty development activities aimed at helping them gain familiarity with these traditions. A number of engineering faculty have participated in seminars on Religion and Ethics in the Professions. Further, in the past five years, our annual School of Engineering faculty meeting has routinely asked faculty to talk about the ramifications of the terms "Catholic" and "Marianist" on our mission in engineering. Finally, our 2005 Strategic Plan offers a very connected vision to the mission defined by our religious traditions. In addition, many (not all) department heads have participated in retreats focused on "hiring for mission," and many have since been working on bringing in new faculty who choose the University of Dayton because of the opportunity to connect their passion to our mission. It is exciting also that we are now rethinking our School of Engineering promotion and tenure document to officially credit our faculty for connecting to mission.

With respect to students, we have found that many, if not a majority of our students are seeking to find deeper meaning in their profession. In our mechanical engineering program, from our Introduction to Mechanical Engineering course to our junior/senior seminars and capstone design courses, students are reminded of the "greater good" role of their profession. Such connection is made primarily by applying engineering problems to societal context. While not all accept this role, those who seek to learn more can do so through mentoring from faculty who they recognize as more connected to our religious traditions. Truthfully, at present, the primary means by which students are able to connect their passion with their professional identity is through this mentoring.

Integration of Knowledge and Problem Solving across
Disciplinary Boundaries

The Catholic notion of unity of knowledge, which derives from the belief that all things come from God and thus all knowledge represents a special revelation from God, embodies the importance of learning across disciplines. We have examples of such integration, but not nearly to the degree that it could be. Our EGR 103—Engineering Innovation I course routinely asks students to think of the connection between a product they are "inventing" and its impact on society (market) and ethics.[18] In recent semesters, our junior/senior capstone design courses have been completely integrated. Civil, computer, electrical, and mechanical engineering students (and others) routinely team on industrial sponsored projects, bringing with them the expertise from their respective disciplines. In the context of these courses, the design teams are required to consider seriously marketing, social, cultural, ethical, and environmental impacts and influences. A design decision analysis is used to evaluate variable designs from these perspectives and ultimately to help pick the best design.[19]

A second means for integration has been through the context in which the knowledge learned in a course may be applied. We offer only two brief examples of what this might mean. Our EGR 202—Engineering Thermodynamics course has involved students in the development of new hybrid energy automobiles with energy harvesting of the waste heat for productive use, the design of solar driven steam power generation systems, and the design of a net-zero energy residence over the past few years. In doing so, their grasp of thermodynamics was improved, but moreover their vision of the need to work on and solve problems aimed at improving energy effectiveness in society was expanded. Some students later indicated that the project helped them for the first time to connect their heart to engineering. Our MEE 312L—Materials course for several years has involved all mechanical engineering students in a project oriented to the evaluation of building materials for cheap, cleaner, and more efficient wood stoves in the developing world. As an introduction to this project, students are presented an orientation to the culture and needs of the poor in the developing world. Ideally then, they are asked to connect engineering knowledge, cultural knowledge, and Catholic social justice ideas.

Outside of engineering, we require our students to take a three-course cluster in themes related to: values, technology, and society; women and culture; cross-cultural, Marianist thought; and the Catholic intellectual tradition, among others. For a cluster, students must take three courses connected by the theme from different domains of knowledge (such as religion, philosophy, history, social science, or the arts). These clusters ideally help students connect knowledge across disciplines. Unfortunately, these clusters have not been looked at fondly by our students. However, there have been some recent experiments of connected engineering/humanities courses that have succeeded. A Civil Engineering/Arts faculty–created course called Constructing Civilization has been well received. In the 2004–2005 academic year, engineering faculty teamed with faculty from religion, science, social science, and business on a course focused on the issue of global warming. Presently, engineering faculty are involved in two multidisciplinary course developments. First is an initiative focused on the theme of Perspectives of Cities. This multidisciplinary initiative involves faculty from engineering, business, social science, the humanities, and our Center for Leadership in Community. This cluster incorporates five courses, each involving multidisciplinary teaching and all seeking to draw connections between the disciplines. It is particularly exciting that engineering is involved in this initiative. Second, is an initiative entitled By Design. This course, to be team taught by engineering and religious studies faculty members, utilizes the design process to evaluate problems in engineering ethics. Both of these latter two initiatives include capstone design projects, requiring connection of knowledge across disciplines and oriented toward some notion of the common good.

A final exceptional example of curriculum related to the unity of knowledge is our junior/senior Professional Development seminar. This seminar exposes students to professional job opportunities, international influences, leadership and teamwork, professional service obligations, and a variety of non-engineering influences on one's career.

While many other examples exist, the truth is that our curriculum still tends to be silo oriented—with separate classes and separate learning experiences. Further, there has been little thought about the developmental path of our students and the connectedness between the different years in the curriculum.

Building and Educating for Community

Among faculty, community building requires faculty who can cross the boundaries existing between the School of Engineering and the rest of the university. We believe that the interaction between our faculty and those outside of engineering is very unusual. In the department of Mechanical and Aerospace Engineering, almost all of our faculty have been involved in institutional curriculum development, academic governance, learning and living, and more. Our New Faculty Orientation Program offers new faculty an early opportunity to do this as well. Further, all new engineering faculty are asked to be involved in a year-long multidisciplinary Teaching Fellows program in their first few years to help them both to learn of various teaching approaches and to meet colleagues outside of engineering. In the past five years, four engineering faculty have also been part of a Humanities Fellows Program which asks for a partnership between an engineering faculty and a humanities faculty, usually in the development and teaching of a multidisciplinary course. Over the same time span, we have had more than a handful of faculty involved in the Catholic Intellectual Tradition Group and four seminars in Religion and Ethics in the Professions/Engineering. Finally, we have a Leadership Development Program that engages future faculty leaders across campus in leadership training from a distinctly Catholic and Marianist perspective. One faculty member from engineering is selected to participate yearly.

With respect to student community, many of our undergraduate students choose UD because of their perception of a strong community here. To create this environment, the university works very hard to get this started well. Our New Student Orientation Program begins the orientation process well before students arrive on campus using a virtual environment and continues with residential life faculty through their first semester on campus.

At a department level, community is also encouraged in all courses. For example, in our MEE 101—Introduction to Mechanical Engineering course, students are reminded of the importance of community and particularly of the importance of teamwork in terms of academic success and in terms of gaining employment downstream. Not surprisingly, employers recruiting our students routinely acknowledge our students' proficiency in teaming and communication.

But the most important aspect of student community is the involvement of students in extracurricular activities. Almost all of our students are involved in service, not because they are required to do so or because employers consider it positively, but because they believe it is important. A strong community among students is also reflected in strong professional student organizations. For example, our American Society of Mechanical Engineers student section has been rated number one in our region in all but two years since 1985, and consistently earned a top three ranking internationally. This merely is representative of the degree of involvement of our students. Our student chapters of professional organizations (including AIAA, AICHE, ASCE, SAMPE, and SAE) also have impressive records in national competitions. Additionally, we have a unique student organization called ETHOS—Engineers in Technical Humanitarian Opportunities for Service Learning. This organization was started by students and connects both to an international internship program in developing countries and local service-learning activities. Finally, engineering students founded the University of Dayton Sustainability Club, which has focused on and succeeded in moving the University of Dayton to more sustainable working and living practices.

At the graduate level, until 2008 there was little noticeable effort to establish community. Since then our Graduate School has been hosting social events. In addition, the establishment of a new masters-degree granting Program in Renewable and Clean Energy in 2009, with a driving, societally connected theme, has provided strong leverage for community development among at least the students now enrolled in this program (currently forty-five). This is an area that still requires improvement.

Community building requires that the boundaries between groups be minimized. Between faculty and students we promote an "open-door" policy, meaning that if students have questions of their faculty, they are encouraged to visit with her or him. In general, this policy is enacted, although the proliferation of e-mail has probably reduced class-related visits by students. Further, a majority of our faculty take very seriously their student advising. In this role, faculty are not merely there to make sure students check the boxes with respect to their degree requirements (we in fact do serve this role), but they are also looking to help students maximize their education and to help them move along paths that best prepare them for life and career goals.

Working as Servant Leaders Striving for Justice Within and Outside of Profession

The importance of educating engineers to understand fully the relevance of issues associated with social justice has become increasingly important as the engineering profession becomes more global. Many engineering schools, including the University of Dayton, have incorporated service learning into their curriculum to best address this theme. Some examples of this are the EPICS® program started at Purdue University and Engineers Without Borders started by Bernard Amadei of the University of Colorado–Boulder. Incorporating service learning into the engineering curriculum has been found to help students develop both technical and nontechnical skills, make connections between classes, develop racial and cultural sensitivity, enhance their commitment to civic responsibility, increase their ethical awareness and awareness of the impact of professional decisions on society and the environment, and see the human side of engineering.[20]

Two prominent manifestations of service learning stand out at the University of Dayton. In our New Engineer Program, a program focusing on the development of first-year students, students solicit as donations used bikes, provide the needed repairs, and ultimately deliver both bike and safety lessons to children from the Dayton community who have no bikes. The second example is our ETHOS program. This program has two main goals. One is to provide full-summer engineering service learning internships in Africa, Asia, and Latin and South America for cottage industries working to help the poor meet basic needs. The other is to contribute to the organizations being served, through first listening in order to assess needs and then acting to meet needs—and explicitly rejecting an approach of telling them what they need. Now, thirty to forty-five students travel internationally working on service projects. These experiences are truly life-changing, and have led to projects that have been integrated into our curriculum. The Materials Laboratory projects referred to earlier are the best example of these.

Faculty must also be models of servant leadership. This is most evident through mission consistent research. At least within our department, that of mechanical and aerospace engineering, much of our research falls under the umbrella of sustainable engineering. This research includes:

- Research and development of cost-effective net zero energy homes
- Industrial energy and waste management optimization
- Energy harvesting of waste heat from aircraft, automobiles, and other sources
- Development of cleaner fuels and more efficient combustors and for aircraft vehicles

Sustainable Engineering Education

The ideal of sustainable engineering education is really embodied in the previous three categories. Sustainability refers to a means of living and practice that insures that future generations can live similarly with resources that are not diminished. As we have perhaps reached the peak for cheap oil, this ideal is not merely an ideal, but a necessity for the future well-being of society. We now offer courses related to green building design, hazardous waste clean-up, design for environment, energy and industrial waste management, and renewable energy systems. Ultimately, these offerings led to the establishment of the program in renewable and clean energy, mentioned above. In addition, also described above, a few faculty have addressed sustainability in the context of required courses. Presently, this theme has found a place in a majority of mechanical engineering courses.

Preparing for Adaptation and Change

No graduate will be able to serve unless they have both knowledge of the global world in which we live and some idea of the action-feedback-response characteristics of society. Presently, our engineering graduates do not have a strong understanding of the modern world. They do not have a sound understanding of the social problems they can influence as engineers. They do not understand the social and cultural impact of technology. They do not understand the need to be adaptive and how to adapt. Instead, they are accustomed to being supplied the knowledge they need. In their engineering courses in particular, they are rarely challenged to be critics of their own solutions or those of others.

MEETING THE GOALS OF A CATHOLIC AND MARIANIST CURRICULUM

Anthony Bright and Clive Dym recently discussed the evolution of engineering education in the United States following the launch of Sputnik in 1957. They suggested that the tenor of the times very much dictated the analytical, science-based approach to engineering education.[21] They asked, "How would engineering education be different today, had the engineering curriculum be posed as a problem in engineering design." Such a design process would require:

- Articulation of properly drawn objectives (What do we want our students to know and do?)
- Articulation of the appropriate and realistic constraints (What restricts our actions?)
- Derivation of the functions that must be performed in order to realize the desired objectives subject to the given constraints (How do we get there?)
- Establishment of the metrics against which the achievement of the objectives can be measured and assessed (How do we know when we have succeeded?)

Bright and Dym further suggest that engineering curricula should be stated as a sum of skills that students are expected to master and a set of experiences in which they will participate. In the past, the focus has been on "What do we want our students to know?" rather than on "What do we want our students to do?"

With this context, we envision an engineering education that achieves distinctiveness through the Catholic and Marianist educational characteristics previously described. A curriculum designed best to address these characteristics must also recognize that constraints exist that may ultimately limit the solution developed, only some of which we consider truly fundamental. But we do not acknowledge constraints placed upon a curriculum by disciplinary silo: in order to achieve our objectives, knowledge silos cannot exist. As such, the following simple and obvious constraints are posed.

(1) Graduates must be employable.
(2) Graduates must be able to complete their education in eight semesters on campus.
(3) Graduates must be prepared to enter graduate school in engineering and other disciplines.
(4) ABET Program requirements must be met.
(5) The delivery of a curriculum to satisfy these objectives must be resource effective.

Constraints 1 and 3 can be addressed through proper consultation with industrial advisory committees and alumni. We certainly believe that if we can more effectively address the engineering implications of Catholic and Marianist theology, then our graduates will indeed be better for both the world and for the companies and organizations hiring our students.

Constraint 2 simply recognizes that the number of credit hours cannot be increased beyond the 132–38 that we presently have. Constraint 4 is easily achievable as ABET now permits universities to more distinctly define what they are.[22] Constraint 5 basically says that at the University of Dayton, in our present environment, we cannot increase the ratio of student credit hours/full-time equivalent faculty. In fact, we have been challenged by our provost to substantially decrease this ratio.

Now that we have identified objectives related to our Catholic and Marianist traditions as well as the constraints under which we must proceed, we can embark on the task of envisioning a functional path for achieving the objectives. The following provides our vision of the foundational basis for developing these objectives in our students throughout our curriculum. This vision is presented to begin the dialogue at the University of Dayton and in Catholic higher education about how to make better connection between engineering education and mission.

Objective 1—*Developing and Living a Passion for the Profession of Engineering*

While a passion for engineering may be present among some faculty, staff, and students, we do not uniformly address this characteristic of our educational goals. We believe that the model of developing a vocational commitment to profession, as provided by our Lilly Foundation–sponsored Program for Christian Leadership, offers a means to achieve this goal, among students, faculty, and staff. This program relies upon education and retreats to offer students knowledge of Catholic Church thinking about vocation. It also provides a forum for personal reflection and sharing, thereby permitting the "heart" to enter into the reason to pursue a profession. Interested faculty and staff could certainly be included in similar programs.

The context of problems in engineering courses is also an important factor in achieving this objective. When the context connects to common-good issues associated with safety and environment and truly helping all in society through the technology we develop, the passion for profession to serve the common good can be nurtured.

In the end, however, there must be a place for vocationally inspired students, faculty, and staff to work. Among students, it is important to help make them aware of the post-degree professional opportunities that best connect to their passion. We must help them to discern which companies and organizations have the value systems that best fit their passion and then provide advice about how to get there. To accomplish this, we should develop career design mentoring opportunities to formalize student thinking in this regard, so that, when they graduate, they have an idea of what they want in their career and how to get there. Moreover, we should work diligently to attract particularly these types of companies and organizations to recruit our students.[23] Among faculty, there must be forums for presenting research/teaching which is inspired by this passion for engineering. Perhaps the establishment of journals which address research crossing disciplinary boundaries and related to helping to support the common good is a necessity.

Objective 2—*Integration of Knowledge and Problem Solving across Disciplinary Boundaries*

The ability to integrate knowledge across disciplinary boundaries arguably is the most important. All other objectives defined build from this one. In order to achieve this objective in our graduates, six functional paths are envisioned to achieve this goal.

First, we need to look at the whole of our curriculum from a developmental perspective. If we say that we want students to develop with respect to the engineering applications of our Catholic and Marianist traditions, then we must continually reinforce this development. Knowledge learned early in the curriculum, both technical and general, should be repeatedly built upon in subsequent courses.

Second, we must have our students learn and solve problems across the curriculum. Most importantly, this objective requires that the learning experience of our students should be pervasive through our engineering courses. Our engineering faculty need to be fully cognizant of the "big picture" goals of the learning that our students are to take away from that part of their education that falls outside of engineering. These big picture goals, which are now more strongly connected to Catholic and Marianist traditions as a result of the recent Marianist Education Working Group initiative, are referred to and reinforced in the engineering curriculum. Faculty development activities certainly will be needed to get our faculty up to speed with respect to these goals and then to provide guidance about how these might be addressed in engineering courses. Examples created by early adopters will undoubtedly be helpful. As described shortly, design may be the best vehicle for delivering on this objective.

Third, engineering faculty need to be far more involved in shaping the general education of all students. Dr. Shirley Jackson, President of Rensalaer Polytechnic University, recently spoke about the lost connection between the "liberal arts" and the "arts" (and humanities). In its inception, she notes, a liberal education included the quadrivium—namely, arithmetic, geometry, astronomy, and music. Later it included the basic sciences of the day, with emphasis on mathematics. Dr. Jackson concludes that: (1) liberal education has generally lost a linkage to the sciences and mathematics, and (2) it has not remained adaptive. Technology, she says, dominates our society, yet most liberally educated people know little about its function and influence on society. Rather than simply offering technology literacy courses for non-engineers, we also need to recognize that engineering notions of system dynamics, feedback, systems, and design have relevance to the general education of all students. Now is the time to begin thinking about educating our populace about the importance and impact of technology in society. Engineering faculty (and students) need to be active players in helping to further shape this education.

Fourth, engineering design—the glue which pulls an engineering education together—must be in everything, and not just in engineering. Clive Dym of Harvey Mudd College has emphasized the place for design in a true liberal education.[24] Dym further suggests that when we rethink engineering education in terms of both knowledge and doing, then design must be everywhere. Learning without applying to new problems helps develop engineers who really do not understand the knowledge we wish them to learn. As Billy Koen has suggested in his book defining engineering, *Discussion of the Method,* engineering is not simply the collection of knowledge, but is the art of solving problems using heuristics that are only occasionally physics and mathematics based.[25] Design activities integrated throughout the curriculum permit development in regard to heuristics of all types (for example, design process, dialogue, persuasive skills, defining *best,* and a host of others).

In a Catholic and Marianist context, design provides the best vehicle for problem solving that relies upon both traditional physics-based engineering knowledge and knowledge normally considered outside of the engineering discipline. The context of the design problems must inevitably ask students to consider obvious common good issues such as environment and safety, but it must also force them to at times ask "If this design is successful, how might it change society? Do I believe in this change? Does the design help the poor or does it further disenfranchise them? Does the design enrich life or will it push more materialism on society?" An engineering faculty truly knowledgeable of the general education "big picture" goals can remind students to consider these and other questions in the context of developing designs.

As much as possible, these design projects must ask students to think about community building among their team and with affected constituencies. They should at least at times ask our students to interact with people who are not the same as themselves. They must work with non-engineering students, industry sponsors, and community organizations. True multidisciplinary experiences will permit students to learn the value of knowledge outside of their discipline in their problem solving.

Design should also be present in the general experience of our students. As a great example, Worcester Polytechnic University students are required to complete a capstone project in the humanities and a project addressing an issue related to technology and society, in addition to an in-major design project.

Fifth, we need to ask students to both identify and solve problems in which they pull together knowledge from multiple disciplines. Problem identification is one of the most important aspects of engineering in professional practice, yet one of the least addressed in universities. Most problems given to students are laid out neatly for them. Even our industrial sponsored design projects begin with a needs statement from the sponsor.

We envision at least two paths for achieving this objective among our students.

- Even when problems are neatly laid out for students, students must be continuously challenged to consider and imagine difficulties inherent to the problem formation. To do this right, they must consider the broader dimensions of problems. They need to be able to say, "OK, we've been asked to solve this problem, but in doing so we will be ignoring issues that likely are important to the development of a better solution." After such analysis, they need to imagine how to frame the problem more holistically and then solve based upon their more holistic definition.

- At the completion of a problem or design, students must be critical evaluators of their solution(s). They must again be challenged to identify problems with their solutions and offer recommendations for solution improvements. They should also learn to be critics of their faculty and their peers, but in a positive, loving sense. We need to help our students gain comfort in sharing and receiving criticism in respectful dialogue with their peers.

Sixth, we need to rethink boundaries between our engineering courses and the rest of the curriculum. Harvey Mudd College, for example, does not produce mechanical, electrical, civil, computer, or chemical engineers. They develop *general* engineers who may have a slight focus in a particular area. But, in general, they have structured their curriculum to emphasize the commonality between disciplines. We have developed an *engineering core* that would provide a common experience for all engineering students through the first two years. Do not all engineers need to understand system dynamics and feedback control systems? If yes, then do we really need each engineering program to address this subject separately? Are there not processes and practices that all engineers do?

As importantly, we need to ask, "Can there be a Catholic engineering thermodynamics?" Mike Sain, an electrical engineering professor at the University of Notre Dame, and Barbara Sain, an assistant professor in theology at St. Thomas University, recently developed a course at Notre Dame drawing analogies between Catholic theology and feedback control systems. In so doing, they learned that such a course offered at least two benefits. One, it helped connect engineering thinking and visualization to the discernment/decision-making process of moral theology. Second, and perhaps most importantly, it helped the students understand that there could be connections between engineering systems analysis and theological analysis.[26]

Objective 3—*Building and Educating for Community*

Given that a vast majority of engineering students are involved in extracurricular and service activities and that teamwork is included in many of our courses, we must understand that community building education is important. At UD, our Center for Leadership in Community has voiced that for *community building* to occur properly, the following characteristics must be present.

- Respect for each individual and their gifts
- A shared vision of both utilitarian goals of the project and teamwork goals
- Time to nurture the social capital of the communities
- Assessment and adaptability
- Skills at engaging in dialogue and deliberation

The starting point in defining how this might be achieved, both curricularly and extracurricularly, is to first define what we mean by community in a Catholic and Marianist context. At the University of Dayton, our Center for Leadership in Community, led by Richard Ferguson and Raymond Fitz, S.M. (former President of the University of Dayton and a former professor of Electrical Engineering), have engaged in thinking about teamwork (community) and leadership from this perspective. In their paper, "Advancing Justice in the City through

Community Building: Themes and Practice Theories Emerging from the Center for Leadership in Community,"[27] which, while focused on community building relative to service learning in the local community, offers generally applicable insight into community and leadership from a Catholic and Marianist perspective. A number of themes are emphasized. Relative to *community building,* the following characteristics are identified.

- Respect for other individuals or groups is implicit. A Catholic viewpoint lifts up the dignity of the individual within community. Respecting others requires an assessment of the assets that the individual or group brings to a team. Ferguson and Fitz suggest that an "assets mapping" be created at the onset of a project. Through this method, each person must be aware of the strengths and knowledge that others may bring to a process. For example, in engineering team activities, some have suggested the importance of the Meyers-Briggs personality survey as a means to do so.[28]

- There must be a shared vision. The vision may be focused not only on the utilitarian output of a project, but also on where the community (team) envisions itself to be in terms of the type of community they have created during the course of a project.

- Mutuality should exist. Individuals working in a functioning community that recognizes the assets of each individual and group understand that they have something to offer to others and have something to learn from others.

- In the early stages of community development, there should be time to nurture the social capital. Relationships matter! Team building activities have value in connecting the community.

- Communities must be adaptive. There must be a periodic evaluation of the technical goals of a problem and in the process being employed. The periodic evaluation should also permit adaptation of the community functioning itself. A community should always be asking, "Are we working together as well as we might be?"

- In order to properly function, a community must have individuals who have the skill to engage in dialogue and deliberation. These skills have not generally been developed in students.

Objective 4—*Working as Servant Leaders Striving for Justice Within and Outside of Profession*

Like many engineering programs, we have a course in engineering ethics. Ethical commitment is best addressed in the context of what we do as engineers—for example, in the context for problems posed to our students and in the context of design problems. We believe our engineering ethics course would be much more effective if it represented a culminating experience for our senior students. In such a course, seniors would employ the knowledge they had acquired in all aspects of their curriculum and, in most cases, through their summer and co-op engineering experiences.

Service is likewise best addressed in the context of problem solving. Service learning experiences in and outside of the engineering curriculum is essential. In addition, our professional development seminars need to more strongly emphasize the obligation to service in our profession. Robert Greenleaf in his very influential 1973 book *Servant as Leader* provided the following definition of a servant leader.

> The servant-leader is servant first. . . . It begins with the natural feeling that one wants to serve, to serve first. Then conscious choice brings one to aspire to lead. . . . The best test, and difficult to administer, is: do those served grow as persons; do they, while being served, become healthier, wiser, freer, more autonomous, more likely themselves to become servants?[29]

These questions also help to provide a metric for assessing the servant leadership abilities of our students.

Further, this type of leadership builds upon the community characteristics described in the previous section with another distinctive feature: leaders must balance inquiry and advocacy to achieve the best solution. This type of leadership understands the importance of facilitating understanding rather than simply winning an argument. Ferguson and Fitz offer further insight about advocacy and inquiry. They say,

> Productive advocacy reveals one's thinking behind points of view and offers examples. Likewise, productive inquiry explores others'

thinking or assumptions and suspends judgment until these are considered. The willingness to engage others' ideas with a sincere desire to understand their points of view is, in our experience, an essential community building skill. Such active listening requires most of us to suspend judgment long enough to listen fully to ideas, explanations, and underlying convictions.[30]

There must also be education about and opportunities to practice servant leadership throughout the curriculum.

Objective 5—*Sustainable Engineering Education*

Sustainability connects to everything we do in educating engineers. It relates to our use of materials, manufacturing, traffic, construction, energy and transport, and design. Sustainability should be a theme integrated into all of our engineering courses, primarily through context. It should also be present in the common educational experiences. Finally, it should be present in our living and working environments. Catholic universities should be leaders in campus energy efficiency and ecological action, not second-class citizens relative to public universities.

Objective 6—*Preparing for Adaptation and Change*

Preparing students to be adaptive requires first that students should better understand the context of the world in which they will work, and particularly those social justice issues relevant to their practice as engineers. Students must particularly understand the impact of their profession on the world today, on all world communities—and particularly the poor and marginalized. Relative to the functioning of the world, students must learn the connection of feedback, action, and change. It also means that we must ask students to at times both learn their discipline and influencing disciplines on their own. Adaptation cannot occur without continuous learning. We do a grave disservice to our students by always providing them the knowledge they need to solve problems. Our students must be both active identifiers of the learning they need and active learners of the knowledge (from multiple disciplines).

Next, design projects throughout the curriculum should be used to provide practice for our students to consider larger issues in their problems. Problems must be posed that require development of potential solutions for a rapidly changing global environment. Students must ask questions such as, "What if oil prices double? How will that affect the solution?"

Moreover, students should learn that design represents an iterative process. Students must be asked to evaluate continuously their solutions and alter and upgrade their ideas as they learn more and as they assess their ideas. They must also continuously assess the processes they are using.

CAN THERE BE A CATHOLIC THERMODYNAMICS?

At the conference panel session led by Father James Heft, describing the most recent incarnation of a seminar on Religion and Ethics in Engineering at the University of Dayton, Brad Duncan, a colleague, friend, and faculty member in the Electrical Engineering Department, offered thoughtful and articulate reflections on the seminar. He further described that the culmination of his participation in the seminar would be an assessment of the degree of embodiment of Catholic and Marianist traditions by engineering faculty at UD. In describing the research he was doing, he acknowledged that while the educational implications of our traditions were important, there could not be a Catholic thermodynamics or circuits. Of course, he was right. Sure, he was right.

This comment, however, has percolated since then. If the characteristics describing the Catholic and Marianist traditions as applied to engineering had true meaning, then it seemed logical to believe that there could indeed be a Catholic thermodynamics course.

If such a course could exist, what would it be?

A Catholic thermodynamics course would minimally rely upon context in problems and projects to draw in common good issues. Addressing these issues could inevitably fuel the search for a passion in the profession. It could also provide a forum for community building in team activities. These we had already imagined.

But other thoughts emerged. The course described above, created by Mike Sain and Barbara Sain, a course that used analogy to draw connections between concepts in theology and engineering feedback and controls, in our opinion, was brilliant. Analogy, we reasoned, could be used to draw connection from engineering to all knowledge and particularly to theology, history, philosophy, economics, and the social sciences, and in a variety of ways. Thermodynamics deals with the identification of systems and the tracking of flows of mass, energy, and entropy into and out of systems. The most difficult aspect of this course for engineering students is associated with identifying the system needed to solve a problem. The greatest harm we do as educators to students is by posing problems where the selection of the system is easy, where the linkages to the world are incorrectly filtered from the problem. Analogy to all of these other disciplines, where systems for analysis are routinely identified, often with much difficulty, could readily be used to help engineering students gain a better sense of how to draw their system boundaries for properly posed engineering problems, for example, where the choice of the system boundary is not so easily imagined and where there is much extraneous information.

A Catholic thermodynamics course could also bring into question the scientific "laws" that we are using, thereby challenging engineering students to question all that they are learning, a necessary characteristic of critical thinking. Einstein's theory of relativity, for example, has proven that the law associated with conservation of mass is at times untrue. Further, despite Einstein's expressed confidence that the second law of thermodynamics (related to the increasing entropy of the universe) was one rule of science that would never be invalidated, the last ten years have seen a proliferation of research at the nanoscale that have yielded such contradictions. Drawing light to these contradictions will help students see a parallel to the *imprecise* theories of history, social science, theology, and philosophy.

A Catholic thermodynamics course could also show how the laws governing thermodynamics have been applied to fields outside of engineering. For example, the first law of thermodynamics has immediate application to electrical and structural engineering, and the second law has been used to describe biological evolution and economic and social science systems. What opportunities exist in other courses to connect engineering knowledge to other fields of knowledge?

A Catholic thermodynamics course could also be based upon the idea that real engineering problems will never be based only on the energy, entropy, and mass balances of a system. Given real problems, students will need to identify the other flows in and out of the system as well as the external system interactions. As an example, consider a design project that might ask students to improve the efficiency of a particular system. Students must imagine the consequences of that efficiency improvement. While engineers have been doing this since the beginning of the industrial revolution, they have rarely sought to glimpse the impact. Energy efficiency improvements have inevitably reduced product cost, thereby permitting people to buy more things. Energy efficiency improvements have also permitted population growth. Our common good goal of saving energy may have actually fueled increased materialism. This visioning is what was called for by conference keynote speaker John Staudenmaier. He suggested that engineers must think about the ramifications of a successful design in the society in which it will be used.[31] They must inevitably rationalize if such a successful design is indeed what they wish for society.

Another element of a Catholic thermodynamics course would be associated with the pedagogy of learning and knowing. How do we know thermodynamics? How do we know theology or philosophy? An approach used by one of us asks students to develop an intuitive understanding of thermodynamics before a mathematical language is used to describe it. We suspect that there easily could be analogies to learning outside of engineering.

A final element of a Catholic thermodynamics course that we can imagine is associated with the notion of best or ideal. A Carnot efficiency defines the best for thermodynamic systems. Can a parallel to best be drawn to theological, philosophical, social, or political systems? This is probably difficult to imagine when the engineering system analysis only considers energy flows. But what happens to "best" once the system is broadened to include flows of things other than mass, energy, and entropy, which are harder to describe mathematically?

We frankly imagine that each engineering course could be thought of similarly. The notions of design, systems, dynamics, energy, and flow all seem ripe for analogy to knowledge outside of engineering.

The real question then would be, "If we did this, would the education of our students be better?" Would we muddy the engineering concepts so much that they would walk away from the course knowing little of these? Or would they better understand these concepts? Would they truly be able to think more broadly and in a more connected way? Would they in the end be better engineers? While much thinking is required to make such a course a reality, we think it worth a try. In the end, we are confident that the answer to the final question is yes.

NOTES

1. David J. O'Brien, *From the Heart of the American Church: Catholic Higher Education and American Culture* (Maryknoll, NY: Orbis Books, 1994).

2. School of Engineering, Proposal for Ph.D. Program, 1973.

3. Phillip Gleason, "American Catholic Higher Education, 1970–1990: The Ideological Context," in *The Secularization of the Academy*, ed. George Marsden, 234–58 (New York: Oxford University Press, 1992).

4. The 2004–2005 Bulletin is far more explicit in identifying the source of UD's educational heritage. The first sentence gives strong life to the Catholic and Marianist traditions: "The University of Dayton is a private, coeducational school founded and directed by the Society of Mary (the Marianists), a Roman Catholic teaching order."

5. David J. O'Brien, *From the Heart of the American Church.*

6. "Characteristics of Marianist Universities: A Resource Paper," 1999, available at http://www.chaminade.edu/marianists/documents/cmu.pdf.

7. Public universities also serve communities but generally not nearly to the degree of engagement at Catholic universities.

8. O'Brien, *From the Heart of the American Church,* 43.

9. Ibid., 105.

10. David J. O'Brien, "A Catholic Future for Higher Education? The State of the Question?" *A Journal of Inquiry and Practice* 1, no. 1 (Sept. 1997): 49.

11. James L. Heft, S.M., "Ethics and Religion in Professional Education: An Interdisciplinary Seminar," in *Enhancing Religious Identity: Best Practices from Catholic Campuses,* ed. John Wilcox and Irene King, 175–99 (Washington DC: Georgetown University Press, 2000).

12. As quoted in O'Brien, *From the Heart of the American Church,* 170.

13. Ibid., 173.

14. Patrick H. Byrne, "The Vocation of the University and Religious Horizons," in *Proceedings of the Center for Catholic Studies: Religious Horizons and the Vocation of a University,* Seton Hall University (2000).

15. Thomas H. Groome, "What Makes a School Catholic?" in *The Contemporary Catholic School: Context, Identity, and Diversity,* ed. Terence McLaughlin, Joseph O'Keefe, S.J., and Bernadette O'Keeffe, 106–24 (London: Falmer Press, 1996).

16. Groome, "What Makes a School Catholic?" 123.

17. Thomas M. Landy, "Environmentalism and Catholic Social Thought: Some Background, Challenges, and Opportunities," *New Theological Review* 9, no. 2 (May 1996): 20–32.

18. K. P. Hallinan, M. Daniels, and S. Safferman, "Balancing the Technical with Social and Ethical: A New First Year Interdisciplinary Design Course," *Technology and Society,* 20, no. 1 (2001): 4–14.

19. Arguably, ABET requires all programs to address these issues in the context of capstone design experiences. However, a checkbox approach aimed at insuring at least minimal consideration of a given issue, as opposed to an integrated one, is most common.

20. E. Tsang, ed., *Projects that Matter: Concepts and Models for Service Learning in Engineering* (Washington DC: American Association for Higher Education, 1999); J. Duffy "Service Learning in a Variety of Engineering Courses," in *Projects that Matter,* ed. Tsang, 75–88; P. H. Wright, *Introduction to Engineering,* 3rd ed. (New York: John Wiley & Sons, 2003); and G. Eisman, "What I Never Learned in Class: Lessons from Community Based Learning," in *Projects that Matter,* ed. Tsang, 13–26.

21. Anthony Bright and Clive L. Dym, "General Engineering at Harvey Mudd: 1957–2003," Proceedings of the 2004 American Society for Engineering Education, Annual Conference and Exposition, St. Louis, MO.

22. The ABET EC-2000 requirements expect that students will demonstrate at least the following outcomes: (i) an ability to apply math and science; (ii) an ability to identify and solve problems; (iii) an ability to design; (iv) an ability to design and conduct experiments; (v) an understanding of ethics; (vi) an ability to communicate; (vi) an ability to function in teams; (viii) an understanding of the global and cultural fabric in which engineering takes place; (ix) an ability to utilize modern tools; and (x) a commitment to life-long learning. Each university has free reign to identify additional outcomes or expand upon those stated by ABET. For example, our present outcome related to understanding ethics is stated instead as: "Our graduates will demonstrate an understanding of and commitment to ethical practice and service inside and outside of their profession."

23. Already, companies and organizations recruiting our students do so because of their perception of differences in our graduates in terms of teamwork, leadership, and ethical commitment. A deepening of these characteristics can only help further in this regard.

24. Clive Dym, "Learning Engineering: Design, Languages, and Experiences," white paper, Harvey Mudd College, 2003.

25. Billy Koen, *Discussion of the Method: Conducting the Engineer's Approach to Problem Solving* (New York: Oxford University Press, 2003).

26. M. Sain and B. Sain, "A Course in Integration: Faith, Engineering, and Feedback" (paper presented at a conference on the Role of Engineering at Catholic Universities, University of Dayton School of Engineering, Dayton, OH, September 22–24, 2005).

27. Richard Ferguson and Raymond Fitz, "Advancing Justice in the City through Community Building: Themes and Practice Theories Emerging from the Center for Leadership in Community," paper delivered at the Catholic Social Thought Across the Curriculum Conference at the University of St. Thomas in St. Paul, MN, October 23–25, 2003.

28. Robert Stone and Nancy Hubing, "Striking a Balance: Bringing Engineering Disciplines Together for a Senior Design Sequence," *Proceedings of the 2002 American Society for Engineering Education Annual Conference & Exposition,* American Society for Engineering Education, 2002.

29. Robert Greenleaf, *The Servant as Leader* (Cambridge, MA: Center for Applied Studies, 1973).

30. Ferguson and Fitz, "Advancing Justice."

31. John Staudenmaier, "Elegant Design Not Enough: Embracing the Tangled 'We' to Critique Technology," paper delivered at a conference on the Role of Engineering at Catholic Universities, University of Dayton School of Engineering, Dayton, OH, September 22–24, 2005.

Chapter Four

Engineering at Santa Clara: Jesuit Values in Silicon Valley

DANIEL A. PITT

This chapter originated as part of a special project at Santa Clara University called Future Directions: Achieving National Prominence as a Catholic, Jesuit University. My original purpose was to issue an invitation to the university community to reflect on the role of engineering at Santa Clara. That we are part of a Catholic, Jesuit university, located specifically in Silicon Valley, is only part of the story. The rest is the particular nature of Santa Clara University, including our history, our mission, and our values. How engineering helps the university achieve its desired prominence is another part of the story. The rest is helping the university achieve its self-defined goals, which all the studies in the special project address.[1]

SANTA CLARA UNIVERSITY

Santa Clara University, the oldest institution of higher learning in California, welcomed its first students in 1851. It was established on

90

the site of the Mission Santa Clara de Asís, the eighth of the original twenty-one California missions. The Mission Church plays an active role in many aspects of campus life—religious and secular—and buildings dating back to 1822 are still in use. Today it enrolls 4,500 undergraduate students and 3,500 graduate and professional students; most undergrads live on or near the campus while almost no grad students do. The School of Engineering offers B.S., M.S., and (a few) Ph.D. degrees, the Leavey School of Business B.Sc. and M.B.A. degrees, the School of Law J.D. and LL.M. degrees plus J.D./M.S.T. and J.D./M.B.A. programs, the School of Education, Counseling Psychology, and Pastoral Ministries M.A. degrees, and the College of Arts and Sciences B.A. and B.S. degrees. As a comprehensive university it ranks number two in the West according to *U.S. News and World Report.*[2]

Three Centers of Distinction serve as major points of interaction among the schools and colleges and between the university and society. The Ignatian Center for Jesuit Education helps the University community realize the evolving role and implications of our Jesuit identity and tradition of faith through justice. Its components engage students in real life, community-based learning experiences both locally and abroad through dozens of formal relationships with community organizations; promote the contemplation of vocation in the sense of finding one's true calling; and organize immersion experiences to provide students, faculty, staff, and alumni with prolonged contact with the poor and marginalized. The Center for Science, Technology, and Society exploits the university's many ties to Silicon Valley to illuminate the interplay of science and technology with culture and society. Among its prominent activities is its support for technology that benefits the less fortunate, through its role in the Technology for Humanity Awards and its Global Social Benefit Incubator for winners of the awards and other social entrepreneurs. The nationally known Markkula Center for Applied Ethics helps people develop strategies for understanding and resolving the ethical dilemmas that confront them and society.

Frequently enrolling first-generation college students, the university strives to prepare and retain all students and has the third highest graduation rate among US master's universities. Approximately 45 percent of undergraduate students and 67 percent of graduate students identify themselves as people of color. All the graduate and professional programs accommodate the working professional, so the graduate demographics mainly reflect those of Silicon Valley's high-tech working population. The university makes student learning its central focus in an educational environment that integrates rigorous inquiry and scholarship, creative imagination, reflective engagement with society, and a commitment to fashioning a more humane and just world. All undergraduate students complete a university-wide core curriculum comprising ethics, liberal arts, and religion. All students benefit by small classes (all taught by professors) and a values-oriented curriculum that educates students for competence, conscience, and compassion. The traditions of Jesuit education—educating the whole person for a life of service—run deep in the university's curricular and cocurricular programs and embrace all faith traditions.

It is worth noting the central role played by El Salvador at Santa Clara. Our Jesuits have maintained longstanding ties with their counterparts at the Jesuit university in San Salvador, Universidad Centroamericana "José Simeón Cañas" (known as the UCA), and supported them personally during the bloody civil war. The president of the UCA, Ignacio Ellacuria, S.J., who was machine-gunned on the lawn of the UCA in 1989 with several others, received an honorary doctorate from Santa Clara in 1987. Santa Clara sends many students and faculty on immersion trips to El Salvador, typically of a week's duration. In addition, Santa Clara runs a study-abroad program there, called Casa de la Solidaridad, for students from all US Jesuit universities, and maintains the infrastructure to support short- and long-term visitors. Santa Clara views El Salvador as an ideal vehicle for applying its principles of learning and contribution. The country remains poor and politically troubled, but the people are full of hope and there are prospects for economic improvement. Our many contacts enable our students to actually make a difference with their efforts. Of course the role of the Catholic Church is strong, as is the people's faith.

With 60 percent of Santa Clara's undergraduate students, the College of Arts and Sciences dominates the campus culture. The college has no graduate program, so it focuses entirely on undergraduate learning, and many university functions, like alumni relations and student life, serve mainly undergraduate students and alumni. Within the college the disciplines of religion and philosophy are viewed as central, and their methods of learning and contributing to society vary greatly from those in engineering. One objective of the original version of this chapter, therefore, was to educate the rest of the campus on the nature of the discipline and profession of engineering. The situation also frames the role of engineering at Santa Clara, requiring it to bridge the dominant campus culture with Silicon Valley.

SILICON VALLEY

Silicon Valley is an unofficial name for the Santa Clara Valley. About an hour south of San Francisco, the Valley runs from San Jose, adjacent to Santa Clara at the foot of San Francisco Bay, to Palo Alto and Menlo Park, a half hour's drive to the north. Named for being the birthplace of the integrated circuit, Silicon Valley has come to embody many areas of high technology. Among the companies whose world-wide headquarters are but a short drive from the campus are Hewlett Packard, Intel, Apple Computer, Google, Yahoo, eBay, National Semiconductor, Oracle, Cisco, Advanced Micro Devices, Xilinx, Electronic Arts, and Genentech. But the Valley represents more than just technology. It represents entrepreneurism, experimentation, challenging the established order, extremely if not excessively hard work, a meritocracy that knows little ethnic discrimination, and, perhaps most importantly, a tolerance of risk and failure unmatched anywhere else in the world. Highly motivated people, especially young engineering and business graduates, still flock to the Valley from all over the world, hoping to make a big difference or just make it big. The Valley has also come to represent vast wealth, often rapidly gained, as well as greed, self-interest, cutthroat competitiveness, and a near-worship of technology for its own sake. The pace is frenetic, the traffic is bad, and the culture is dominated by high-tech. Those who can't hack it, leave, because the cost of living in the Valley is too high for many people who lack the financial rewards of stock options. The median price of a house in Santa Clara County, which runs from San Jose to Palo Alto, is over $700,000. You would be shocked at how modest a house you might get for that money.

THE DISCIPLINE OF ENGINEERING

When Santa Clara's School of Engineering opened its doors in 1912, an engineering education prepared students primarily for the development of civic infrastructure and secondarily for development of technologies derived from the industrial revolution (mechanics and electricity). In the ninety-four years hence, engineering has expanded considerably to become a leading unit in the university and a leader in the region, but its purpose remains to provide the most evident advances in societies and civilizations: sanitation, housing, transportation, communication, production of goods, and tangible vehicles of commerce and economic development.

We view the engineering design process as a triad: We start with a foundation of mathematics and the physical sciences (representing laws of nature we cannot change), seek solutions to human needs and desires, and bound our choices by economic reality. Imperfect, constrained problem solving constitutes the basis of engineering, and the mental discipline to incorporate basic knowledge, design creativity, and the judgment to weigh suboptimal solutions serves an engineer well both within the sphere of professional activity and in other aspects of life. In his special-project paper on ethics and justice, Paul Fitzgerald advocates exposing students to "truly ambiguous and vexingly difficult situations," and while he refers to ethical ones these are the very conditions engineers encounter regularly in their work.[3] In a similar vein, two respected North Carolina educators, Richard Felder (engineering) and Rebecca Brent (education), have recently outlined the necessary progression of students from absolutists who believe in the perfect wisdom of others, to contextual knowers who respond to the complexities of the world by taking responsibility for their own learning.[4]

Because we need to experiment with and develop real systems, engineers face the vexingly difficult task of actually getting things to work. Often the most vexing part is having to sacrifice the ideal aesthetic in the design for the mundane constraints of practicality and cost. Add to that the human difficulties of communicating and understanding design specifications and the daunting challenges of bringing to a successful conclusion projects that are larger (often many times larger) than what one person can tackle, and you have a multidimensional profession of which technology is only a small part. A solution must fall within the social, moral, and sentimental lines drawn by users. This is finally an iterative process, one often marked by conflict, as users are not monolithic and often desire different things. There is also an element of teaching and translating, as many users do not know what is realistic or possible. To be an engineer is to listen to all these voices simultaneously.

For this reason, an engineering education prepares students for many careers (and is especially well suited as preparation for medicine and law). Consider also the many career transitions an engineering graduate can expect throughout the duration of a career, not only within engineering but beyond. Fewer than 50 percent of engineering graduates will practice engineering to the end of their careers. The others will move, temporarily or permanently, into roles of sales, product marketing, customer support, management, or other functions in companies that also do engineering, not to mention those that move into entirely different careers. But the thought process and mental discipline, especially calling for creativity, problem solving, communication, and judgment, remain valuable over a lifetime.

FORMATION OF ENGINEERING STUDENTS

Students typically enter engineering school knowing little or nothing of the profession. They usually like doing things they associate with engineering, such as building things or playing with computers. We doubt that many have considered the contributions engineers make to society, though the evidence surrounds us constantly. Many students, like the public at large, equate engineering with technology. Identical they are not.

We know that engineering suffers from a poor image among the general populace. Dr. William Wulf, President of the National Academy of Engineering, quoted a Harris poll finding that only 2 percent of the respondents associated engineers with the word "invents"; only

3 percent associated them with "creative"; meanwhile, 5 percent associated them with the phrase "train operator."[5] Most students who choose engineering had the benefit of knowing someone in the profession—a relative or friend of the family—though not what they actually do at work.[6] But this kind of connection only perpetuates the lack of diversity in the field. In particular, the public's lack of familiarity with the profession of engineering makes it difficult to recruit and retain women and minorities.

Showing the social relevance of engineering by engaging the students with the community in an effort to define projects that meet real needs is a win-win-win proposition. Involvement in community-based projects at Santa Clara provides motivation to our students and is instrumental in educating them in the conscience and compassion aspects of our university mission. Obviously the community members win by getting access to the free development of a project for which they feel the need. The community members also gain confidence in their own abilities to deal with and express their views on technology.

The third win is the most subtle, but perhaps the most far reaching. In working with the students, the community members learn a little about what engineers do. Little by little, we see an understanding growing within the community that engineering is a creative and people-oriented profession. They also see that women and minorities can be engineers. Over the past three years, 75 percent of the students involved in these projects were female, and 50 percent were members of an ethnic minority underrepresented in engineering. These students were drawn to the opportunity to do something valuable for the community; the opportunity was offered to all students, but appealed most to the women and minority students. (Our engineering school student body during this period was approximately 23 percent female and 15 percent underrepresented minorities.)

So our job, as engineering educators, is to aid their formation as leaders contributing to the creation of a more humane and just world, with engineering skills and judgment dominating their toolbox. We move their thinking from tasks to outcomes; we broaden their perspective to include the real world to which Father General Peter-Hans Kolvenbach refers to when he calls us to "educate the whole person of solidarity for the real world."[7] We place engineering in a context that includes the other disciplines of the university, but we do not do this well enough. One of the treasured benefits of Santa Clara engineering is the inclusion of ethics, religion, and liberal arts for engineers, but there are two problems. One is that most of our students do not have time to take enough of these courses. They do enjoy the community experiences we offer them both locally and internationally, but their basic requirements are heavy. To maintain our accreditation we have to cover a lot of ground in engineering and most students cannot afford to take five years for a B.S. degree (though the national average for engineering is 4.8 years). The other problem is that our students seem to mentally separate engineering and the university core. Thus we have to do a better job in our engineering courses demonstrating the relevance of the Santa Clara core to our students.

The Jesuit values get introduced overtly in the human-needs side of the triad that defines the engineering discipline and in the judgment required to find a solution under the constraints of the three sides. We encourage students to understand "the conflict between the urge to do the right thing and an understanding of the complexities of what the right thing is."[8] We want students to consider needs more than desires, to pursue both "the good of each and the good of all" to the extent they can.[9] Both engineering and ethics share the objective of utility (social, economic, technological), and from a theological viewpoint both are rooted in the tension between what is and what ought to be. The objective nature of engineering provides little or no room for self-deception, and sustained exposure to an unyielding reality can lead to a kind of self-discipline that is conducive to a discerning faith. Thus engineering can reinforce concepts taught in the university core.

Note that many engineers cannot know or predict how their inventions and designs will be put to use. How would the person designing error-correcting codes for loss-prone transmission media know that they would one day enable farmers in remote villages in India to share a cellular phone to ascertain crop prices? Or a drug dealer in Medellin to arrange invisibly a shipment of cocaine from Cartagena to (ironically) Corpus Christi? Through our senior design projects and faculty research we make every effort to confront students with opportunities to advance social justice through engineering, and we already have a reputation for community-based learning, as with our student projects in El Salvador, described below. We know "it is not sufficient to give students direct experiences that make them want to work for justice"[10] and "all too often, students are rushed into the field to make justice happen, without sufficiently rigorous intellectual inquiry into what justice means and how its conditions ought to be fulfilled,"[11] but it is unacceptable for us to deny them these experiences. They must, however, be accompanied by the "whole person" education we claim to offer, which also incorporates and informs the research of our teaching-scholar faculty.

THE BLESSING AND CURSE OF SILICON VALLEY

The School of Engineering at Santa Clara occupies an often coveted location in Silicon Valley. We are embedded in the midst of opportunity, innovation (in technology, business formation, employee practices, funding methods, core competencies), and—the Valley's most distinctive feature—acceptance of risk and failure. We are surrounded by success, but also by excess. No place is closer to a meritocracy (evidenced by cultural diversity), but a pure one still eludes us; it works pretty well for ambitious men and Asians, less so for women, Hispanics, African Americans, and devoted parents. Engineering students benefit by learning from those who have succeeded, and failed, sometimes in spectacular fashion. They taste the chance to contribute to innovations that the whole world knows about, and they are afflicted (positively) by the zeal for entrepreneurism.

Despite its fame and benefits, the Valley presents the school with challenges. The needs of employers change rapidly, and sometimes irrationally, and the time-constant of change is shorter than the time students go through college. So the job market they find when they graduate could be quite different from the one that existed when they enrolled. The fortunes of companies rise and fall overnight, and employees move often between companies, making it difficult for the school to maintain long-term relationships with industrial partners.

Topics that need to be learned by undergraduates as well as working professionals change a lot faster than the lifetimes of tenured faculty. And the needs of graduate students in general are pointedly career-focused, if not purely job-focused, and they are not seeking formation.

Indeed, one of our most difficult challenges is understanding how to respond to the needs of the working professional, in his or her thirties or forties, often of foreign birth, for graduate education. It would be hard to articulate what makes their graduate education distinctly Santa Claran. And it is tempting to dismiss their need as being alien to Santa Clara's mission of basic education and character formation for resident undergraduate students, but we are not a university located in rural mid-America. This is our community, and we cannot turn our back on it. How in good conscience can we exhort our undergraduate students to engage in lifelong learning and inculcate in them a desire to do so, and then when our graduate students are doing just this, simply say "Not in my backyard"? Moreover, this community needs us for our values and as a refuge from the stress and excess the Valley dynamics engender.

We cannot settle for anything less than an engineering program worthy of our location; such is a condition of the university's achieving its ambitions. To draw the talented professionals and families in Silicon Valley (and students who aspire to be so) to the school we have to offer them a high-caliber engineering education in appropriate traditional and emerging fields. To challenge them to become leaders of competence, conscience, and compassion in the process, we align our curriculum to better reflect our values, we invoke community-based learning and other tools that foster growth of character in learning, we reflect our values in our own actions daily, and we hire faculty and staff with all this in mind. Thus we infuse the Jesuit values into the broader community, one that typically does not receive such influence.

THE OPPORTUNITY

The opportunity before us is to leverage the School of Engineering's commitment to advance the mission of Santa Clara with concrete means that will benefit undergraduate and graduate students, the school, and the university. Having a great, Jesuit-values-infused engineering program that is well integrated into the university will benefit all parties involved; it brings the values to a previously underserved community and brings their engagement and resources to the university. Here are some ways we are trying to do this.

1. Defining a broad, common engineering core that is not discipline specific and that concentrates on the fundamentals of the profession. This eliminates inefficiencies across the departments and allows us time to relate engineering to the intellectual challenges of ambiguity, complexity, and justice that are common across the university. The core could also include community-based learning for all students in engineering at all levels.

2. Emphasizing interdisciplinary programs within engineering and between engineering and other schools at Santa Clara. Indeed, we seek a variety of means of integrating engineering students and faculty with their counterparts across the university. This keeps the school at the forefront and distinctive academically, it strengthens other parts of the university, it applies engineering to many problems whose solution advances the university mission, and it projects the university into the community. We have started to do this with our Center for Nanostructures and are investigating how to incorporate bioengineering into our program. Other interdisciplinary themes emerging from the faculty are environmental sustainability, mechatronics and robotics, information assurance and security, and social science studies of technology.

3. Introducing system design projects in every year of the engineering curriculum, and including business topics as part of the system. The role of engineers in Silicon Valley will increasingly require greater understanding of business questions and customer needs, as routine engineering moves offshore. Every senior design project will eventually entail consultation with business students and a component of the presentation that addresses the business case for the project. We expect that stronger ties with the business school will eventually pervade our undergraduate and graduate programs, as will ties with local industries and the entrepreneurial community. We have a new student organization, the Engineering and Business Alliance, with

both undergraduate and graduate students working together. Doing all this with an eye toward solutions applicable to the economies of the developing world contributes to the university's goal of healing economic disparities.

4. Increasing diversity in our engineering student body not simply through focused recruiting but through a recasting of the profession in terms of achievable impact rather than isolated activity, and through outreach activities where appropriate. Our experience, and that of others, has been that minorities and women are more attracted to engineering if they see what it can do for their communities. Santa Clara engineering is blessed with the fifth highest percentage of female faculty of any accredited engineering school in the country,[12] at nearly three times the national average, and we strive to provide these faculty with support to enable them to serve as ambassadors for the profession and leading advocates for the recasting. We are barely above the national average in percentage of African American and Hispanic faculty (at 5 percent each); we would like to double that. We are ranked 35 out of 350 engineering schools in the country in our percentage of undergraduate women, but frankly the competition is not that stiff, and all of us can do better.

5. Exploiting the university's Centers of Distinction to give a uniquely Santa Claran flavor to the challenges of our time. The engagements of our students and faculty with the Markkula Center for Applied Ethics and the Center for Science, Technology, and Society encourage our students to confront issues that we see in unpleasant headlines all too often.

6. Leveraging the university's ties to El Salvador as our response to preparing students for the challenges of globalization and the globalized engineering profession. In addition to immersion programs, engineering students can now take courses with students from the UCA via live video and whiteboard hookups that operate in both directions. Today, this immersion program is administered through Santa Clara's Casa Educational Network, a new study abroad program emphasizing accompaniment with the local community, academics, community, and spirituality. One team developed a more cost-efficient and sustainable construction method for Northern Ghana using catenary arches and stabilized laterite clay bricks. Another

team designed a filtration and distribution system for Pajarillos, a rural community in Honduras. Their filtration system was implemented into Pajarillos' existing water system in March 2010. Another developed a solar-powered water purification system capable of supplying enough clean water for an average household throughout the third world. Another designed a water purification system based upon a vapor compression distillation process. Most striking today is that these international projects now encompass all disciplines.

To achieve our goals and still enable our students to graduate in four years is not easy. However, if we concentrate on arming them with the fundamentals that will last them a lifetime, and leave more of the deeper, discipline-specific studies for graduate school, we will have succeeded. Too much of what engineers are taught at the traditional engineering schools, whose emphasis is narrowly focused on technology, becomes obsolete within ten years of graduation. A Santa Clara engineering education strives to yield graduates who can make informed choices about what to learn throughout their careers. We want our graduates to be well-rounded professionals, aware of the impact their work and careers have on the world. We want them to have the career agility the profession demands, especially in Silicon Valley. And we hope that their conduct, at work, with their families, and in their communities, marks them as products of Santa Clara Engineering for the admiration of all, most notably in Silicon Valley.

NOTES

The author wishes to thank Mark Aschheim, George Fegan, Silvia Figueira, Tim Healy, Brian McNelis, Leigh Star, David Tauck, and Alex Zecevic for their thoughtful comments on the original paper, and Ruth Davis for her contributions to this one.

1. The papers representing these studies, some of which we refer to, can be accessed at http://www.scu.edu/strategicplan/futuredirections/index.cfm.

2. *US News and World Report,* "America's Best Colleges 2011."

3. Paul Fitzgerald, "Ethics and Justice as Integrating Factors in a Santa Clara Education," December 14, 2004, available at http://www.scu.edu/strategicplan/futuredirections/themes/ethics.cfm.

4. Richard Felder and Rebecca Brent, "The Intellectual Development of Science and Engineering Students: Part 1, Models and Challenges," *Journal of Engineering Education* 93 (October 2004): 269–77.

5. American Association of Engineering Societies, "American Perspectives on Engineers and Engineering," a Harris Poll Pilot Study conducted for AAES (Washington, DC: AAES, 1998).

6. In a first-year engineering class in fall 2004 at Santa Clara, fewer than 7 percent of the students had no family or close family friends in engineering.

7. Peter-Hans Kolvenbach, "The Service of Faith and the Promotion of Justice in American Jesuit Higher Education," public lecture, Santa Clara Lectures, Santa Clara University, October 6, 2000, vol. 7, no. 1, available at http://www.scu.edu/news/attachments/kolvenbach_speech.html.

8. Alan Wolfe, "The Intellectual Advantages of a Roman Catholic Education," *Chronicle of Higher Education,* May 31, 2002.

9. Paul Fitzgerald, "Ethics and Justice as Integrating Factors in a Santa Clara Education."

10. Mark Ravizza, "The Mission of Santa Clara as a Catholic, Jesuit University in a Globalizing World," December 14, 2004, available at http://www.scu.edu/strategicplan/futuredirections/themes/mission.cfm.

11. Alan Wolfe, "The Intellectual Advantages of a Roman Catholic Education."

12. Accreditation Board for Engineering and Technology, 2005.

Chapter Five

A Systems View of Time-dependent Ethical Decisions

HAMID A. RAFIZADEH AND

BRAD J. KALLENBERG

SYSTEMS VIEW OF ETHICAL CHALLENGES

Every ethical situation has a "system" characteristic with a group of human and nonhuman elements linked in a variety of interactions and interdependencies. The system allows the elements to act in part or as a whole towards achieving a spectrum of goals, objectives, or ends.[1] The systems view asserts that any local and bipolar understanding of an ethical situation would be deficient as it would neglect certain interactions and interdependencies as well as overlook differing orientations of agents towards different goals and objectives. The purpose of this paper is to highlight the need for a systems-based view of ethics.

Systems thinking is not, of course, a panacea. It is one thing to have an intellect informed about the Good and quite another to have a volition reliably oriented to choose the Good. This latter issue is of central concern to Christian ethics but is not the focus of the present chapter. What is of concern to us is the usefulness of systems thinking for revealing the Good. This paper details five "systems thinking" principles by analyzing a specific example of employee behavior under top management pressure.

From a systems point of view, ethical considerations are tightly intertwined with operational considerations and thus ought to attend to general features observed in system dynamics. For example, the observation that in operations people have a poor understanding of system dynamics would equally apply to ethical considerations. First, people see a system as an aggregation of components and not as interactions evolving over time. Second, people fail to recognize the significance of time delay between action and response because they tend to have a linear, here and now, view of causality.[2] Our thesis is that systems thinking can help sensitize us to the dangers of overlooking the time delay between actions and their consequences.

Each individual is located in a variety of multilayered systems. For example, as a worker, the average person stands in a web of relationships with co-workers, subordinates, and superiors. The pattern of interactions between them constitutes the system called the corporation. The corporation is not merely the aggregate of all the people who work there. It is the people plus the habits of their multilayered interactions over time. The corporation itself is a part of the local business community. As such it stands in various political and economic relationships not only to its own employees but also to other corporations, the consumers, the state, and so on.

But an individual also exists in a living community. As such, he or she is enmeshed in a web of relationships including nearby neighbors. The neighborhood interlocks with other neighborhoods, and together these comprise the political community. The pattern of relationships within this neighborhood and between neighborhoods makes up the corporate persona of this municipality. Municipalities relate to each other within economic realities and legal and political strictures within the county, state, and nation.

So then, each individual is simultaneously a worker, a neighbor, a family member, and so on. Each system is multilayered within itself and is intercalated with other systems. The layering is not predefined and can be variously dependent on where one draws a layer's boundary and what one includes within it. Nonetheless, there are two features common to all systems. First, for all individuals the system is a nested structure. Second, some arbitrary boundary cutting across the layers is taken by a given individual as a "boundary of significance." The human attention and expenditure of resources primarily reside within the boundary of significance. For most individuals the boundary of significance encompasses the home and workplace. A similar layering exists for the corporation. It defines the boundary of significance through setting objectives and allocating budgets. We argue that ethics requires intentional expansion of one's boundary of significance to view the surrounding things in increasingly holistic ways.

For example, few people are conscious of the extent of their connection with the biosphere.[3] Let us start with the typical person's automobile, a popular component within the boundary of significance, transporting the individual between work and home. How much attention does the individual pay to the exhaust emitted from the automobile? There is a good chance that an awareness of the exhaust manifests once every two years when the automobile must pass its emissions test. Otherwise, the individual rarely if ever "sees" the flow of the exhaust gases into the atmosphere and even more rarely wonders about its effect on the atmosphere. Now consider the gas furnace in the house. Does the individual ever pay attention to gases going up the chimney and into the atmosphere? Perhaps in some winters, receiving the gas company's bill, he or she may complain about the cost and the furnace gas consumption, but the furnace's gaseous emissions remain unnoticed. Another element within the individual's boundary of significance is "electricity." To the individual it is a clean form of energy. His mind does not venture to the layer in which the power plant generating the electricity is located. Only there would the huge column of gaseous emissions rising into the atmosphere be noticeable.

How do automobiles, gas furnaces, and electricity demonstrate the deficiencies of the typical moral vision? They do so by showing that the person focused on home and workplace does not see the consequences of gaseous emissions in layers beyond the home and workplace. Moreover, the typical ethical individual develops little understanding of the time delay between emissions from automobiles, gas furnaces, and electric power plants and the accumulation of the society's emission gases in the atmosphere. The emissions within the individual's layer of significance seem small and innocent. Yet, in today's conditions, at the final global layer the gaseous emissions accumulate at the rate of 22 billion tons per year.[4] Assuming a human population of 6 billion, there is an accumulation of about four tons of gaseous waste per year per person. Any accumulation at the rate of four tons per year would have quickly caught the eye of the individual if it happened within the boundary of significance or even in the next layer or two. But this is taking place at the last layer, that is, at a global level. Can we argue that since this is the system's last layer, it may not be significant for the individual even if it accumulates at the rate of four tons per year per person? Such a conclusion is entirely unwarranted, because the buildup of gaseous emissions in the atmosphere can result in global warming. Here we arrive at the problem of "knowledge flow" across system layers. The knowledge relevant to the first layer may not reside there but in other layers. For example, while the words "global warming" may reach the individual, the implications of global warming do not. He or she would regularly dismiss as irrelevant the notion that earth's average temperature may increase by a degree or two.[5] The individual sees noticeably higher temperature variability from morning to noon every day. The knowledge that fails to reach the individual is that the "average temperature" of earth controls the weather pattern and the weather pattern controls the output of agricultural products. A few degrees change in average global temperature can produce weather patterns that wreak havoc in the world's food supply.[6] With deficient food, the items lying within an individual's boundary of significance, namely the home and workplace, can face immense danger. Yet such a systems-based linkage of the home, workplace, and the globe remains beyond the embrace of the individual's awareness. From the system dynamics point of view, this is not a surprise but a fact of human life—the inability to see time-delayed effects of one's actions within the context of the whole system. Driving between home and workplace, the use of a gas furnace to heat the home, and the electricity to light the house and operate various gadgets, produce byproducts that accumulate globally to endanger the individual's boundary of significance decades later.

ETHICAL DYNAMICS—ANALYSIS OF
A SIMPLE EXAMPLE

The inability to see time-delayed relationship between action and response prevents the individual from comprehending the ethical dilemma of altering the atmosphere and its properties. Even if the adverse effect on agriculture does not materialize, annual dumping of four tons of anything in the global backyard is significant enough to be noticed, but it rarely is. Though the issue of gaseous emissions highlights ethical human behavior with respect to long-term global issues, it is not an ordinary daily activity. To analyze more specific system dynamics of ethical considerations we will rely on a more mundane example.

In a business situation in 1984 in a regulated energy company, the engineer/manager in charge of the forecasting department prepared an energy consumption forecast to be presented to the regulatory agency to set the prices the company charged customers. The forecasting department developed a complex econometric model that projected a 4 percent growth in energy consumption. When the results were presented to the company CEO, he stood firm that the forecast was too high. With tens of adjustable parameters in the econometric model, the company's forecasting expert could reset some to create a lower rate of growth. The first parametric adjustments produced a forecasted growth rate of 2.5 percent. The CEO remained adamant that it was still too high. Another round of parameter adjustments produced a forecasted 1.5 percent growth. The CEO was pleased but added that he would like to see a forecast with 1 percent growth. With some grumbling the forecasting expert adjusted a few more parameters and the company forecast exhibited a 1 percent growth in energy consumption. A thick report capturing the company's econometric model was prepared and submitted to the regulatory agency.

The first ethical question concerns "parameter adjustments." Is it wiser to set the parameters at what the forecasting expert deems appropriate or at what the CEO declares proper? The positions taken by the CEO and the forecasting expert intersect at the historical data. The choice of action differs by the choice of pointing to different layers of the system. The CEO could point to the pre-1981 growth numbers which were at 1 percent or less and declare the 3 to 4 percent trend of 1981–1984 as an aberration, while the forecaster could point at the econometric model and claim that the higher trend was not an aberration but a result of underlying developments in the economy. Which position would be more correct?

We claim that the CEO is acting legally but unwisely because he is choosing to ignore the best guess of the expert practitioner. This leads to an error that could have been avoided by expanding the range of one's system awareness.

We offer a remedy in five steps. The first step of systems thinking is that the parameters must be set in a context broad enough to view all of the relevant parameters. Therefore, the choice of parameters ought to be expanded to include the full range of logically possible parameters. In their conversation, not all parameters of the econometric model were well defined. For the sake of clarification of argument we will assume that the choices for each parameter could be represented by a normal distribution as shown in figure 5.1. Arguments based on statistics

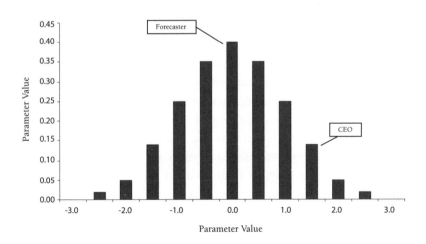

Figure 5.1 Which Choice is More Ethical?

would claim the parameter value should be set at the most probable. But statistical arguments do not overrule subjective considerations. For example, in flipping a coin the forecaster would go with 50 percent chance of heads or tails, while an expert gambler could go with the 100 percent feel of three consecutive heads. The statistical considerations cannot rule against the gambler's sense of events. They remain at a stalemate. Which position is more correct, the one purely based on most probable considerations or a mix of statistics and subjective considerations? Can the same argument be made for positions taken by the forecaster and the CEO?

The second step is to understand the system as an embodiment of a "portfolio of objectives." Since all systems are layered and intercalated with other systems, one must specify some range or other for analyzing the system(s). (To define the economic system as what's in my wallet is too narrow. To examine my household budget in light of its impact on India's soft-drink industry may be too broad.) Within a given range, each system has an ordered set of priorities or objectives. This set is called its "portfolio of objectives."

The larger a system gets and the more layers are considered, the more distributed the portfolio of objectives becomes. For the forecaster, the key issue within a narrowly restricted range is the reliability and accuracy of the forecast. But we can surmise that in the CEO's outlook parameters such as profitability are the primary objective of the portfolio. In effect the two were engaged in a conversation in which each took a position relative to a system quite different from the other. There was no attempt at identifying the definition of the system, its layers, and the boundary of significance. Their miscommunication was further complicated by the fact that the portfolio of objectives is not only relative to a given range, it is also hierarchical. Even if the CEO and forecaster had presumed the same objectives, it is clear that for the CEO profitability sits higher than accuracy and reliability of a forecasted aspect of corporate operations. What role does systems thinking play in determining the ranking of a portfolio of objectives?

From the CEO's point of view, profitability is inversely linked to the forecasted growth potential. Consider the regulated company's total expenses, including a reasonable profit, as Total Cost. This has to be divided by the average consumption over the period under consideration to set the price charged to the customers. The average consumption, however, varies by the growth rate. The price the company can charge customers is thus determined by the relationship:

$$\text{Price} = \frac{\text{Total Cost}}{\text{Average} \left[C_o \prod_n (1 + \alpha)^n \right]}$$

where α is the growth rate, n the number of years under consideration for rate setting, usually three to four years, and C_o a constant factor determined by historical consumption. The pi operator is a product operator. This equation simply states that price can be forecasted from the growth rate, and that the forecasted price is the product of growth in year 1, 2, 3, and so on.

From the CEO's point of view, a low forecasted growth rate would set the price high, improve profitability and reduce risk, while a high forecasted growth rate would do the opposite. While in appearance the CEO-forecaster dialogue is centered on reliability of the forecast, it in fact engages two totally different objectives simultaneously. The forecaster remains unaware of the CEO's profitability objective and can only assume the CEO's concerns emanate from the pre-1981 historical growth rate of about 1 percent. Is a dialogue that does not reveal the *entire* "portfolio of objectives" ethical? Or, more generically, at what level of knowledge deficiency do the ethical considerations become irrelevant, overwhelmed by poor knowledge?

As we have seen, systems thinking reveals that the CEO-forecaster conversation artificially restricts the portfolio of objectives under consideration. But their conversation is skewed because the forecaster mistakenly imagines the CEO as the only significant other in the conversation. The third step in systems approach, therefore, involves the forecaster expanding his or her vision to include the silent partners. What appears to the forecaster to be a local and binary interaction actually involves another key player, the regulatory agency. The forecast that the company prepares and sends to the regulatory agency is reviewed and scrutinized by the regulatory agency's forecasting experts. In setting the price, the regulatory agency can declare the company's forecasted growth rate low and adjust it to a higher number deemed more appropriate by the agency's forecasters. It is the regulatory agency's function to protect the customers from corporate overcharging and high prices. The regulatory structure exists primarily on the assumption that the energy company will try its best to maximize profitability and charge the customers the highest price possible. Given such an arrangement, how does it modify our view of ethical behavior of the CEO and the company forecaster? Are the regulatory agency's forecasting experts and commissioners as ethically responsible, or perhaps even more responsible, than the energy company actors? Unlike the corporation, in principle, the agency should not have any internal profitability mandate, and its primary objective should be to set fair prices.

The ethically interesting aspect of this three-element view of the system is that often the regulatory agency chooses to accept *unquestioningly*[7] the company's forecast as valid. Is it possible that the regulatory agency and its own analysts do not comprehend the company's econometric model or do not understand the impact of growth rate on prices charged to customers? The truth is simpler than that. The regulatory agency allows overcollection by the company because it can allocate part of the overcollection to its own favorite social engineering experiments in the form of "energy efficiency projects" *which it has no other way of funding!* (In short, kickbacks make the world go 'round.) If the company is held, for example, to a consumption growth of 4 percent, the regulatory agency will have no leverage on the company to force it to take on the agency's "social engineering experiments." However, if the regulatory agency allows the company to collect more from customers under the assumption of a 1 percent consumption growth, then it can negotiate for a share of the over-collection towards the agency's energy efficiency programs.

Is any of this unethical? Not in the sense of being illegal. However, insofar as no party has troubled to look at the wider system, it is the economic equivalent of dumping pollution into the air by means of an inefficient furnace. After all, the regulatory agency does not take from customers for personal or organizational gain but for experiments that if successful may provide value to all customers. Similarly, from the CEO's point of view the negotiations with the agency are simply to set the level of profitability of the company. The more profitable the company, the more it will be able to serve the needs of the customers more efficiently at a higher level of financial stability.

From the systems point of view the ethical situation that started with the "choice of forecast parameters" ended up with "extra resources taken from the customers" and allocated to corporate profitability and social engineering. The different choices of the system's boundary of significance make it collectively defective.

For our present purposes, the difficulty in "seeing the whole picture" is a function of three things. First, the engineer/manager, the forecasting expert, and the CEO lack a systems thinking education. During their education, none learns the basics of defining and seeing the interactions and interdependencies of the workplace in systems terms. Second, no training program at the workplace corrects the lack of "systems thinking training." In fact the workplace's fragmentation and segmentation ensures that no one at the lower ranks would develop a systems view of the corporate operations. The executives at the top develop a biased view of the system tilted towards corporate profitability within a layer of significance defined by the corporation.

Third, even if the universities taught systems thinking to students and even if corporations adopted a code of conduct to train employees in systems thinking, system dynamics tells us the few that control corporate resources, namely the executives, *define* the system. The CEO would have no problem bypassing a systems-conscious forecaster by simply assigning the preparation of the forecast to a consultant who would gladly produce a 1 percent growth forecast in return for its consulting fee. Because of this power structure, the ethical character of the entire system would largely depend on the behavior of its resource controller, namely the corporate executives. Nevertheless, it is our hope that systems thinking education and training have the potential to substantially reduce the possibilities for unethical behavior by the executives.

SEEKING THE SOLUTION IN THE WISDOM OF CHRISTIANITY

It is wrongheaded to imagine blithely that, moving on separate paths, systems thinking and Christian ethics will inevitably converge on common morality. A systems thinking approach that is disconnected from Christianity will inevitably narrate the world as an economic story of individuals and their aggregates (corporations) who compete against each other in a zero-sum game of survival of the fittest. Such opposition between players clearly does *not* tell the Christian story of creation and redemption. The resulting portfolios of objectives would be desperately disordered. The proper order will not emerge by simply staring at the data. Rather, what is needed as a fourth step is a theological vision of the system.

In other words, what is missing in the present picture is a Good common to all (because it comes as a gift), one that provides orientation for all the proximate goods and in light of which all other goods must be ranked. Pope John Paul II wrote in *Laborem exercens* that competition for profit is not the root of all evil—an unfair charge commonly leveled against Christian thought. Rather, keeping evil from taking root requires setting up the system in such a way that it (1) maximizes profit without destroying others, (2) maximizes creativity and internal goods for workers and practitioners, and (3) maximizes the Common Good for all, including those who are outside the walls of the corporation (not only the consumer, but also, for example, those who live down river from the plant).[8] This vision of the Common Good views the system at its widest scope. It asks us to see the business world as contributing to the Common Good rather than as epitomizing survival of the fittest.

The fifth step in taking a systems approach is to ask the question of time delay. Not only are systems "spatially" layered and intercalated, they and their component parts (in our case, human beings) are extended through time. It took the forecaster, Hamid Rafizadeh, about twenty years to comprehend the system dynamics of parameter adjustments he made in the econometric model. The revelation took place after a few years of studying and teaching systems thinking. Even then, it was accidentally triggered in the aftermath of a classroom setting in 2005 where the professor asked, "Have any of you been pressured by your boss to do something not morally right?" To the forecaster, now a student in the class, the 1984 actions in the 1984 time frame looked professionally and ethically above board. But looking at the events of 1984 with the knowledge base of 2005, which included a deeper understanding of the corporate and regulatory agency relationship, he was no longer sure. Could he have been able to stand up to the CEO's demands if he were as aware of the corporate system then as he is now?[9] But surely this is a moot point. For, the forecaster at that time simply could not see it.

Did his ignorance then make him innocent of any ethical wrong-doing? Maybe yes, maybe no. St. Thomas Aquinas explains that ignorance relates to culpability in a variety of ways. Blameless ignorance he calls "antecedent ignorance"—the kind of ignorance that is genuine and simply precedes the agent's action. But there are three other kinds of ignorance. *Concomitant ignorance* describes the case in which had the agent known better, the agent would still have chosen the present course of action. If knowing better makes no difference to the course of action, then the person acting is just as blameworthy either way. The other kinds of ignorance fall under the category of "consequent ignorance," so called because it is a state of ignorance adopted after or for the sake of the action taken. For example, *affected ignorance* describes the case in which the agent chose to avoid knowing for the sake of having an excuse! The other form of consequent ignorance is called *ignorance of evil choice*. It describes the case in which the agent was genuinely ignorant, but is so because of a flaw in his or her character. For example, perhaps the agent did not take the time to read the warning or was too lazy to study the appropriate data. In all these latter cases, ignorance does *not* spare the agent from being morally blameworthy.[10]

We see, then, that Aquinas does not give a simple answer. Yet in our story, the forecaster's ignorance seems genuinely of the "antecedent" variety. That is to say, the forecaster's ignorance is blameless in that it is not a product of some character defect in the forecaster. That being so, the forecaster is most likely *not* culpable for the economic sins committed by the power company against the consumers on the basis of the forecaster's report. Granted, the forecaster signs off on a growth rate that seems low, but still allowable within the parameters of the model employed. What the forecaster does not know as a young professional in 1984 is the hidden scheme of the CEO. But two decades later, it dawns on the forecaster that the CEO had ulterior motives for setting the growth rate lower than predicted.

The clever reader will have discerned that the "time dependence" given by the two examples is clearly not equivalent. In the case of the automobile and furnace emissions, global warming is the natural accumulation of greenhouse gases over time. Because the time of the accumulation is very long—much longer than twenty years—the significance of the single consumer is trivial. The impact of a single consumer goes unnoticed. In the second case, "time dependence" refers to the maturation of the forecaster over a twenty year period. The forecaster's gaining of awareness is not simply the accumulation of information; otherwise the time delay could be countered by intensive uploading of information on the front end of the forecaster's career. But growth of insight can never be simplified to mere acquisition of information. Rather, his coming to know was time consuming because in addition to information, the forecaster needed time to develop the skill of sorting what he was seeing. As Aristotle noted, cultivating the habits of moral vision necessarily takes time. While twenty years is the blink of the planet's eye, twenty years is a very long time in the life of an individual.

Educators and mentors of engineers are players in a system that includes, for example, forecasters for power companies. The leverage they exert on students and underlings falls short of being able to orient volition—only God can do that! But they do have a role to play in shaping the outlook of those they teach and guide. It has been said that if a burglar is breaking into the house, it is too late to *start* lifting weights! Fortunately, students are not yet facing burglars, corrupt CEOs, or crooked regulatory agencies. In the meantime, educators must urge students to begin lifting moral barbells. In our view, the five steps of systems thinking is a helpful conditioning program that promises to shorten the time delay between now and the day when students become engineers who see and act holistically when faced with ethical decisions.

NOTES

1. Howard Eisner, *Essentials of Project and Systems Engineering Management,* 2nd ed. (New York: Wiley, 2002), 5; A. Kossiakoff and W. N. Sweet, *Systems Engineering Principles and Practice* (New York: Wiley, 2003), 3; Harold

Kerzner, *Project Management: A Systems Approach to Planning, Scheduling, and Controlling,* 7th ed. (New York: Wiley, 2001) 70; Benjamin S. Blanchard, *System Engineering Management,* 3rd ed. (New York: Wiley, 2004) 6; A. P. Sage and J. E. Armstrong, *Introduction to Systems Engineering* (New York: Wiley, 2000), 5–6.

2. John D. Sterman, *Business Dynamics: Systems Thinking and Modeling for a Complex World* (Boston: McGraw-Hill, 2000), 21–22.

3. David Ehrenfeld, "The Cow Tipping Point," *Harper's Magazine* 305, no. 1829 (2002): 13–20.

4. Robert E. Morrison, *Global Climate Change,* Congressional Research Service Issue Brief, Report No. 1B89005, March 2, 1989, 4.

5. Less that 9°F separates us from the last Ice Age; "Evidence of Global Warming," 2005, http://www.ecobridge.org/content/g_evd.htm, last accessed October 2005.

6. H. M. Kaiser and T. E. Drennen, eds., *Agricultural Dimensions of Global Climate Change* (Delray Beach, FL: St. Lucie Press, 1993).

7. A case of what Thomas Aquinas called consequent ignorance (specifically, *affected* ignorance).

8. *Laborem exercens* names as "the indirect employer" all stakeholders and consumers. These are all each of the corporate system and thus have both rights *and duties.*

9. Even if he were capable of systems thinking as he is now, as a lone individual would he not be powerless to influence the CEO when tens of millions of dollars of corporate profits are on the line? He would be thrown out and replaced with a more accommodating forecaster or bypassed for a consultant who would create a forecast desired by the CEO. This brings us to the question of the power dynamics of the system. The powerful person at the top cannot be swayed by actions of a lone individual. The fate of the corporate whistleblowers is well known. See C. Fred Alford, "Whistle-Blowers: How Much We Can Learn from Them Depends on How Much We Can Give Up," *American Behavioral Scientist* 43, no. 2 (October 1999): 264–77; and M. P. Glazer and P. M. Glazer, "On the Trail of Courageous Behavior," *Sociological Inquiry* 69, no. 2 (Spring 1999): 276–95.

10. See St. Thomas Aquinas, *Summa Theologica,* trans. Fathers of the English Dominican Province (New York: Christian Classics, 1981), I–II.6.8.

Chapter Six

Pursuing Dialogue between Theologians and Engineers

JAME SCHAEFER AND PAUL C. HEIDEBRECHT

WHY THEOLOGIANS NEED TO REFLECT ON TECHNOLOGY

While many reasons should compel theologians to reflect on technology in general and some technologies in particular, five major goals sparked interest in exploring the theology-technology connection at Marquette University: (1) to facilitate the unification of student knowledge and skills; (2) to specify the distinctions between theology and technological disciplines; (3) to discern and demonstrate ways of constructively relating theology and technology; (4) to serve as examples of theology-technology collaboration for students as they complete their formal educations and enter their professions; and (5) to formulate more plausible discourse about God in relation to the world.

Facilitating the Unification of Students' Knowledge and Skills

Departmentalizing disciplines continues to be the norm for most universities, especially at research institutions of higher learning. Professors tend to speak in mutually incomprehensible languages, leaving students to fend for themselves without the knowledge and skills they need to relate the disciplines to one another. To mitigate these inadequacies in a student's educational experience, some universities are moving toward cores of common studies that link a plurality of disciplines considered essential for an undergraduate education. One or more theology courses are included in the core curricula of Catholic universities. Theology and other core courses constitute "the heart" of Marquette University's academic program that is designed to give students "a variety of ways to examine, engage and evaluate the world" and a "cross-training" of the human brain that "reflects the values of the Jesuit educational tradition and Marquette's mission to form lifelong learners committed to excellence, faith, leadership and service."[1] Marquette is also making strides toward relating disciplines through interdisciplinary courses and a "senior experience" within which the disciplines address a particular topic.[2]

Offering students who are seeking degrees in technological fields a formal opportunity to relate theology to their major academic concentration is vital to the type of education Marquette and other Catholic universities strive to provide. However, helping students to bridge their theological and technological studies requires theologians to have some basic knowledge about the phenomenon of technology in general and various technologies in particular. Unless theologians obtain this knowledge, they will not be able to reflect in substantive depth to aid their students in the quest for a more comprehensive educational experience.

Specifying the Distinctions between Theology and Technological Disciplines

Key to the quest for unifying knowledge is distinguishing between technology and Catholic theology, the primary religion examined at Catholic universities. Defining technological and theological disciplines requires identifying their distinct data, methods, purviews, and limitations. Students should understand that each discipline has its own methods and contents and is not qualified to do the work of the other.

Identifying the data, methods, and limitations of technology is especially important for theologians so they can direct their students in reflections that are relevant, meaningful, and helpful. The astounding technological advancements that have been made in recent years attest to human ingenuity, while the adverse effects from some of these technologies, especially on those least able to protect themselves, affect how theologians think about the human person in relation to other persons individually and collectively, other species, ecological systems, and the biosphere of Earth.

Identifying the content and methods of theology is especially crucial in light of the literal approach to sacred texts by some sects that allege to follow the Abrahamic traditions. At Catholic universities, students learn to take a critical approach to these texts, as proffered by the Pontifical Biblical Commission of the Roman Catholic Church,[3] thereby recognizing the contexts of the times in which they were written, the circumstances that gave rise to these texts, and the authors' various understandings of the world. Students in theology courses are given an opportunity to appreciate how biblical writers and theologians have over the centuries tried to write meaningfully, relevantly, and fruitfully about their relationship to God. Engaging in theological discourse that is meaningful, relevant, and fruitful requires students to recognize the scientific and technological age in which they are enmeshed today. Key to this theological endeavor is realizing that discourse about God is always inadequate in light of the subject and is open to change when attempting to think about God and how God relates to the world in which humans are integral constituents.

Also significant to the task of defining theology is recognizing the multiplicity of religious traditions within the student population at Catholic universities. While Catholicism may be the privileged tradition on our campuses, the three Abrahamic traditions share some basic beliefs that can be highlighted when bridging the disciplines of theology and various technologies. Judaism, Christianity, and Islam profess belief in one creator and sustainer, think about the universe as totally dependent upon God, and reflect on the responsibility humans have to God for making and executing their decisions. These beliefs provide foundations from which to reflect critically on the research, development, and application of technology in general, as well as on particular technologies. Teachings beyond these shared principles of faith can be appropriated and applied for their deepest possible meaning and application. It is essential for Catholic students to become familiar with social teachings on technology that have been issued by popes and councils of bishops, the *Catholic Catechism,* and critical reflections by theologians on these teachings. Students who profess other religious faiths can become informed about Catholic teachings as part of their educational experience on our campuses, but opportunities for them to become familiar with pertinent teachings of their own traditions and to apply them to their projects practically and theoretically should be made available in the ecumenical and interfaith spirit established by Vatican II.

Identification of the disciplinary purviews and limitations of theology and technology should lead to recognizing where their expertise lies when addressing a technological issue and the unique contributions that each discipline can make when engaged in dialogue. Instead of thinking that one discipline has dominance over the other, they will be viewed as complementary when approaching mutual issues from their distinct perspectives.

Discerning and Demonstrating Ways of Constructively Relating Theology and Technology

While theology and technology are distinguishable from one another, professors need to explore how they can be related constructively on issues at their boundaries. The most obvious way is for theology to provide ethical direction on the research, development, deployment, and use of technology, as physicist and theologian Ian Barbour has aptly demonstrated.[4] There are, however, other ways that may be less obvious. Among these are: (1) answering ultimate "why" and moral "how" questions about human functioning in the world; (2) thinking more cogently about the human person; (3) constructing a model for technologists from religious teachings; (4) providing a horizon of wisdom within which scientific and technological achievements can be viewed and employed; and (5) affirming technology as a type of human labor. We discuss each briefly.

Informed by general knowledge of technologies, Catholic theologians can be equipped to identify religious foundations for developing and using technology by answering ultimate "why" and moral "how" questions about human functioning in the world as informed by their relationship to God. Answering these questions is especially crucial since some technologies affect humans as well as other species and ecological systems directly and indirectly in negative as well as positive ways. Informed Catholic theologians should be able to point to principles of faith that cause technologists to pause and reflect on the ramifications of projects they are contemplating, planning, and executing. Among these principles are maintaining the dignity of the human person, respecting life, having special concern for the poor and future generations, seeking the common good, and valuing the physical world as God's creation.[5] Why and how humans should function in ways that reflect these principles are questions that theologians should help students answer. The "why" questions may be easier to answer from a theological perspective than the "how," since technical issues are usually complicated and defy an easy solution. However, knowledge of real and projected effects of technologies should facilitate answering these questions when tapping into traditional Catholic thinking, especially teachings about the virtue of prudence and the other moral virtues that prudence informs as discussed below.

Theological reflection on technologies can also help theology professors engage in more cogent theological discourse about the human person, the more-than-human others that constitute Earth, and the planet as a whole because of their relationship to God as their mutual creator and sustainer in existence. That humans evolved from and with other species over billions of years of cosmological-biological development must be factored realistically into how humans think about themselves in relation to one another, other species, ecological systems, the biosphere, and the cosmos. A more humble view of *homo sapiens* surfaces—especially when realizing that humans are radically dependent upon other species, the air, the land, and water for material sustenance—as well as a sense of responsibility to God for how technology is developed and used can be explored. The specialty of theological anthropology is ripe for reflection and especially fruitful when addressing technology in general and various technologies in particular.

Awareness of near- and long-term ramifications of technologies contributes to the construction of a model for human behavior that orients technologists toward their relationship with God. Many possible exemplars of people in relation to God can be appropriated from the Catholic theological tradition, though some from biblical, patristic, and medieval teachings need updating to reflect our contemporary understanding of the world. For example, retrieving the teachings of St. Thomas Aquinas on the integrity of the world and the moral virtues is particularly helpful for modeling the technologist as a virtuous cooperator.[6] From Aquinas's perspective, developing the habit of being prudent will assure a step-by-step approach to discerning the best possible technology to apply in the best possible location to meet intended needs with minimal adverse effects now and into the future.[7] Informed by the results of this discernment process, technologists can avoid excess and waste by developing the virtue of temperance,[8] and they can mitigate adverse effects on human and ecological systems by developing the virtue of justice.[9] Development of the virtue of fortitude[10] will help technologists follow the best possible plan, consistently act professionally, and avoid swerving from this plan for selfish or other nonprofessional reasons. A morally grounded technologist in the Catholic tradition will be motivated by the theological virtue of love, which ultimately, according to Aquinas, is the desire for eternal happiness in the presence of God.[11] In addition to Aquinas's model and others that are retrievable from the Catholic theological tradition, students who belong to other religious communities should be encouraged to search for and evaluate models that their traditions offer.

When informed at least generally by technological advancements, theologians can provide a horizon of wisdom within which these advancements are viewed and employed. In 1998, the late Pope John Paul II urged scientists and technologists "to continue their efforts without ever abandoning the *sapiential* horizon within which scientific and technological achievements are wedded to the philosophical and ethical values which are the distinctive and indelible mark of the human person."[12] Catholic universities can promote this horizon by learning from the wisdom literature of the Bible how to live orderly lives that are conducive to a right relationship with God.[13] Reflections on the wisdom literature by prominent theologians throughout the centuries can also be explored and, where necessary, reconstructed to reflect the technological times in which we currently live.

Theology informed by technology can also affirm the enterprise of technological development, which is often viewed ambivalently,[14] and can caution against the proliferation of technologies that may threaten the health and well-being of people, ecological systems, and the planet. While some technologies have enhanced and prolonged human life and well-being, others have spewed persistent toxicants into the air, on the land, and into waterways; produced stockpiles of nuclear wastes that await isolation from the biosphere for thousands of years; genetically enhanced humans, animals, and plants without sufficient attention to present and future ramifications; replicated nonhuman animals and anticipate replicating humans; and facilitated access to information and ideologies that threaten society. Theological reflection on technology as a manifestation of human labor, long lauded in Catholic social teachings since the end of the nineteenth century,[15] can affirm the quest to develop technologies that are helpful to people, especially the poor and powerless now and into the future, so that they can secure the necessities of life as they journey toward their goal of eternal happiness with God. Of course, theological reflection must also be extended to encompass the effects that technologies have on the more-than-human environment with its diverse species and ecological systems, which, with humans, constitute God's continuing creation through the evolutionary process.

Thus, theologians who are knowledgeable about specific technologies can facilitate student reflection on and understanding of the usefulness and functionality of theology in relation to technology. That theology and technology can be related constructively out of mutual respect for the contributions they make to a more comprehensive approach to issues at their boundaries is an invaluable lesson for students to grasp. That theology can be helpful to students who are pursuing engineering and other technological degrees when they are considering the research, development, and use of technologies can be amply demonstrated by identifying and applying principles of religious faith to projects they are contemplating.

Serving as Examples of Theology-Technology Collaboration for Students.

Constructive contact between theology and technology professors can expand the horizons of both, clarify their roles at a Catholic university, and better equip them to facilitate the goal of providing a comprehensive educational experience for their students. Professorial collaboration can also serve as an example for students to carry into their professional lives so they can be open to a range of possibilities and thereby avoid a myopic view of projects in which they are engaged. Acknowledging the other's discipline, respecting it, and showing how religious faith can relate constructively to technological studies constitutes a great service that professors in Catholic universities can and should provide for their students.

Formulating More Plausible Discourse about God in Relation to the World

Theological reflection on technology can spur more cogent thinking about God in relation to the world. As noted above, biblical and post-biblical discourse about God over the centuries reflects the contexts of the inspired writers' times, the circumstances that prompted their writing, and their understandings of the world.[16] Extensive efforts have been underway for over thirty years to engage explicitly in theological discourse that is informed by contemporary scientific findings, as evidenced by the burgeoning literature, conferences, and courses offered on Catholic and other university campuses. A major aim of these efforts is to reconstruct discourse about God and the human person so it is relevant to our times, meaningful for personal and community reflection, and helpful when considering how humans should act. Underlying this effort is the realization that theology is ongoing by those who believe in God and who wish to express their faith in ways that are plausible in light of extant knowledge about the world. Being informed by contemporary scientific findings that underlie technologies tends to facilitate more plausible theological discourse.

While theological reflection on technology in general and various types of technologies in particular opens to thinking about the human relation to other people, other species, natural systems, and the universe, as noted above, reflection on the phenomenon of technology and scientific findings that underlie and facilitate technological endeavors also open to more plausible and meaningful discourse about God in relation to the world. The most basic findings by scientists indicate that the universe has developed cosmologically over nearly fourteen billion years and produced at least one planet on which biological life has emerged with one species eventually capable of producing astounding technologies. When reflected upon, these findings prompt more cogent ways of thinking about God. God can be understood as generous by endowing the world with the capacity to become the universe that scientists study today and within which technologies are developed and advanced. God can be understood as empowering the world to self-organize in increasing complexity to a point in time where humans have emerged to develop highly complex technologies. God can be understood as freedom-giving by not interfering in or interrupting the natural processes out of which the universe is developing through God's sustaining action. God can be understood as patient while at least one species opens up to and responds to God's self-communication on how its members should function in the world. God can be understood as calling the entire universe of constituents to completion and human constituents to orient their efforts toward fulfillment in God's eternal life.[17] By reflecting on God informed by the scientific findings that undergird technology and the technological experiences that prompt further scientific theories and testing, theologians will be continuing the quest to talk about God from the context of our time and scientific understanding of the world.

These and other rationales should prompt theologians to reflect on technology in general and various technologies in particular. The depth of these reflections will depend, of course, on the specialty of the theologian, though the subspecialties of systematic theology and ethics opens most appropriately to the scholarly pursuit of the technology-theology relationship. As already indicated, a key impetus to this pursuit is examining human agency in the world, which is the focus of theological anthropology.

SEMINAR ON THEOLOGY, TECHNOLOGY, AND ETHICS

One way to encourage theological reflection on technology in general and specific technologies is to include this topic among the standard array of topics covered during the formal education of theologians. In recent years, the list of special topics in systematic theology and Christian ethics at Marquette that merit the focused attention of graduate students has been dominated by traditional or highly specialized theological issues.[18] Some of the special topics seminars provide the opportunity to examine themes that relate to technology—for example, ecological ethics, economic ethics, and war and peace. However, for reasons discussed above, devoting an entire seminar to the topic of technology seemed increasingly appropriate in order to sustain discussion of the following questions: (1) Does theology have a role to play in response to the development of new technologies and in the use of established technologies? If so, what is that role and how should it be played? (2) To what extent do the sparse reflections on the theology-technology-ethics relationship point to fruitful ways of engaging theologians now and in the near future? (3) What resources can the Catholic and other Christian traditions draw upon to advance theological reflection on technology and norms for guiding its development and use? Are any of these resources particularly helpful for thinking about human agency to guide human behavior in the world?

Offered for the first time in the fall of 2004, the theology, technology, and ethics seminar attracted seven graduate students—two masters level and five doctoral level students—which is a typically sized graduate class in the Department of Theology at Marquette. Both the experience of designing this seminar and teaching it for the first time evidenced the need for greater reflection on technology within all theological traditions. Nevertheless, this effort constituted a modest but promising first step toward the larger agenda of encouraging dialogue and collaboration between professors of theology and technology.

The assigned readings for the seminar were divided into three broad categories in order to provide an overview of historical, theological, and ethical perspectives on technology in general. This shift in perspectives was intended to provide a transition from a more descriptive to more prescriptive analysis of the relationship between theological thought and technology, although, to a certain extent, all readings included a combination of both types of analysis.

We began by reading David Noble's *The Religion of Technology*, as well as excerpts from Susan White's *Christian Worship and Technological Change*, Jay Newman's *Religion and Technology*, and David Hopper's *Technology, Theology, and the Idea of Progress*.[19] These readings raised a number of crucial questions that were subsequently tracked throughout the seminar, including: (1) How is technology defined? (2) Is the definition broad or narrow? (3) What is the underlying attitude toward technology? (4) Does the author lean toward viewing technology as a threat or a sign of progress? (5) To what extent are models and norms provided for human agency in relation to technological development and use? and (6) What is the religious perspective or the religious tradition that informs the author's scholarship, and how sound is the author's grasp of that tradition?

While David Noble's portrayal of the Christian tradition is selective of sources that are not fully representative of that tradition, his book provided a provocative starting point for the course. Noble argues that the history of technology is properly understood as religious history, since technological development accelerated whenever it was invested with religious significance. The unifying thread throughout this history is the identification of technology with transcendence, an identification that is characteristically Christian: "Christianity alone blurred the distinction and bridged the divide between the human and the divine. Only here did salvation come to signify the restoration of [human]kind to its original God-likeness."[20] Noble also thinks this identification is problematic.

> On the deeper cultural level . . . technologies have not met basic human needs because, at the bottom, they have never really been about meeting them. They have been aimed rather at the loftier

goal of transcending such mortal concerns altogether. In such an ideological context, inspired more by prophets than by profits, the needs neither of mortals nor of the earth they inhabit are of any enduring consequence. And it is here that the religion of technology can rightly be considered a menace.[21]

Despite the hostility toward Christianity that Noble portrays in this book, his thesis underscores the need for a background in theology to fully understand technology and for the kind of dialogue that was facilitated in the Marquette seminar. Located as we are *within* the Christian tradition, we would go beyond Noble to argue that, in addition to explaining the past to provide guidance for human interaction with technology in the future, a theological exploration of technology must strive to be prescriptive as well as descriptive.[22]

Theological perspectives were provided by Willem Drees, Carl Mitcham, Jacques Ellul, and Robert John Russell, a collection of lectures by Philip Hefner, and essays by Paul Tillich.[23] As this list of authors might suggest, it is difficult to find places where contemporary theologians address the topic of technology in a sustained manner. Certainly Ellul's work must be grappled with in any course of this type, although his background was in critical social thought rather than theology. Credit must be given to Mitcham for his work in editing the only and now dated anthology on theology and technology, but he is a philosopher not a theologian. Tillich might appear to be the one exception, although his interest in technology grows out of, and is clearly secondary to, his broader interest in the relationship between theology and culture. In a parallel manner, the interest of Drees, Russell, and Hefner in technology grows out of their significant work in the emerging field of religion and science.[24] This by no means diminishes the potential value of the theological resources provided by these authors, but it does highlight the difficulty in making the topic of technology a priority in theological circles.

Seminar discussions in this section tended to focus on each author's understanding of the human person in relation to the physical world and God, and in this regard Ellul and Hefner represented two extremes in the spectrum of authors considered. Ellul's emphasis on the autonomy of technological systems and ways of thinking appears to limit human agency, a perspective that most likely flows from his Reformed theology's teachings. However, far from suggesting that Christians are powerless, Ellul's emphasis on the autonomy of technical progress demonstrates his view that mastering progress is ultimately an ethical and spiritual problem. Indeed, Ellul insists that "scientists and technologists" heed the gospel message:

> The challenge to our very existence posed by science and technology today can only be met on the basis of a spiritual renewal, on the discovery of a new foundation for human life . . . namely, on the basis of the choice of non-power and on the practice of liberty.[25]

Hefner emphasizes a dynamic understanding of the human person, characterized as a "created co-creator," and considers technology as part of the process of "human becoming." Put even more strongly: "Technology is either pointless in the long run, or it is an expression of the fundamental self-transcending reality of God."[26] He argues that technology can extend rather than diminish human agency, and a theological tradition can provide the necessary perspective for directing humans to proper ends. While Hefner's larger project certainly merits consideration, both he and Tillich fail to provide sufficient normative guidance. They appear to accept the world in general, along with the technological milieu within which we live, as neutral. Our technological culture is what it is, Hefner and Tillich indicate, and we need to change the way we orient ourselves to it.[27] Because engineers and other technologists are intimately involved in the development of technologies, they are better placed than theologians to recognize that particular technologies are far from neutral. Technologies not only embody the values of their human creators; they also encourage the adoption of particular values, or at least shape the existing values of their users. Thus, technologists can inform theological reflection so it is more relevant and meaningful for developers and users of technologies.

Ethical perspectives were provided by selections from Hans Jonas's *The Imperative of Responsibility,* Albert Borgmann's *Power Failure,* and by Murray Jardine's recent book, *The Making and Unmaking of Techno-logical Society.*[28] Jonas, Borgmann, and Jardine are philosophers who are concerned with the relationship between religion and technology. More importantly, they share a common concern for praxis and agree on the need to identify a way of embodying values and convictions in our actions and practices. Jonas argues that the unprecedented power and reach of modern technology, a phenomenon that Borgmann refers to as the "device paradigm," is a primary reason why we find it so hard to connect values to actions in the world. We simply lack the ability to grasp the full impact of our technologically mediated actions. Jardine agrees, arguing that the "profound moral crisis" in the West is the re-sult of our "inability to make moral sense of our technological capa-bilities."[29] In other words, we are unable to make moral sense of our capacity to change our environment through technology because our moral framework assumes that this environment is part of an unchang-ing natural order. He goes on to argue that the Christian tradition is partially to blame for this inability, and that a transformed Christian worldview provides the solution.

Like theologians concerned with Christian ethics, engineers and other technologists would find Jonas's and Jardine's intellectual pro-posals rather unsatisfying. Conversely, the work of Albert Borgmann holds great promise for finding common ground, since his primary concern is to transform our practices, not our ideas. In short, Borg-mann suggests that there are times when Christian convictions lead to practices that challenge the technological practices of contemporary Western societies. Yet help is needed from all angles to negotiate the shape and implement these "focal" practices in order to create commu-nities that will both restrain and redeem technology.

While the seminar students took turns leading the discussions on the assigned readings and provided helpful background information on the authors, their role became even more significant during the final part of the seminar. A full class was devoted to the presentation of each student's research project in which manifestations of technology were engaged theologically and ethically by drawing on the resources of a particular theological tradition. Each of these specific technologies was identified early in the semester in order to minimize overlap and maximize the sharing of sources. The technologies included both cutting-edge and mature topics: genetic engineering, sex-change technology, technology for care of the terminally ill, the pharmaceutical industry, television, automobiles, and music-making technology.

Fortuitous for this seminar was the fact that the participants came from a variety of denominational backgrounds including Roman Catholic, Eastern Orthodox, Christian Reformed, Evangelical Protestant, and Mennonite. Their diverse perspectives were most clearly articulated during the presentation of research projects. In every case students were forced to choose theological conversation partners who had not directly addressed the topic at hand, much less technology in general. This proved at times to be a challenge, although in each case it was confirmed that theological resources are available to spur further reflection on issues related to technology. The efforts of major theologians such as James Gustafson, Karl Rahner, Stanley Grenz, and John Howard Yoder, along with Catholic social teachings and statements from the Reformed Catechism proved helpful when applied to the technological issues researched by the students. However, they recognized the need to continue to peruse their traditions for promising ideas that can aid theological reflection in a technological age.

Thus, the relative dearth of theological reflection on the topic of technology in the scholarly literature provided both a challenge and an opportunity. The seminar challenged the students to locate readings appropriate to their research projects, and it provided an opportunity for them to work creatively and constructively on these projects. From these efforts, the students recognized many possibilities for future research as they considered topics for their master's theses and doctoral dissertations.

MOVING TO THEOLOGY-ENGINEERING COLLABORATION

Lessons learned from conducting the theology, technology, and ethics seminar help considerably toward shaping a future course. One that relates theology, engineering, and ethics is warranted at Marquette and other Catholic institutions that boast strong engineering curricula. The ideal course will be team-taught and offered for either theology or engineering credit at the undergraduate or graduate level. A team-taught course will provide an opportunity to engage both engineering and theology faculty on the larger issues at the boundaries of their disciplines and on issues pertinent to specific projects. Having both faculty in the classroom for the duration of the course will allow interchange of perspectives and demonstration of how each discipline is supposed to function when their data, methods, purviews, and limitations are respected.[30] This exchange of perspectives will be fortified by offering a maximum of opportunities to the theology and engineering students to demonstrate how the two disciplines will function when addressing engineering projects.

A future theology-engineering seminar will consist of four sections. The first will focus on a historical overview of the relationship between engineering and theology. In the second, foundational theological and ethical constructs in the Catholic and wider Christian traditions will be identified from which to address engineering in general. The third section will be dedicated to examining one engineering project in sufficient depth and reflecting theologically on that project from the foundations and ethics identified in the first section to guide the development of the project. In the culminating section, students will be teamed in pairs to research and represent either theology or engineering when looking at a particular engineering project, present their cases in class, and write a research-reflection paper integrating the two disciplines on the project.

Reading sources will vary in the first section of the seminar. It will begin with the introductory section of Barbour's *Ethics in an Age of Technology* to set the stage for characterizing how students think about technology at this point and to discern the ways in which subsequent authors think about technology. Until a comprehensive text is available to provide a historical overview of the theology-technology relationship,[31] some of the readings from the theology, technology, and ethics seminar will have to suffice. A critical view of Noble's *The Religion of Technology* will at least provide an opportunity to identify other sources in the Christian tradition that need to be incorporated to understand more fully the complexities of the technology-theology relationship while also recognizing the dangers in the chilling strains of Christian thinking that Noble exposes so vividly. Excerpts from White's *Christian Worship and Technological Change* will be particularly useful when examining how particular technologies have been shaped by cultural and specifically religious values as well as how some technologies have influenced theological discourse and practices. Selections from Newman's *Religion and Technology,* Hopper's *Technology, Theology, and the Idea of Progress,* John Staudenmaier's *Technology's Storytellers,* and John Brooke and Geoffrey Cantor's *Reconstructing Nature*[32] will also be helpful in providing a historical understanding of technological development. This section will close with an article by Staudenmaier in which recent trends in studying the history of technology are overviewed.[33] Throughout this section, dialogue between the professors is crucial to help students recognize key ideas underlying the theology-engineering relationship historically and determine any significance these ideas have today.

In the second section of the seminar, foundational theological and ethical constructs in the Catholic and wider Christian traditions to which the students ascribe will be identified in order to address engineering in general. Theological approaches taken by Ellul and Tillich, who have pioneered interest in the relationship between theology and technology, will be examined along with more recently published texts by Drees, Russell, and Hefner.[34] Russell and Hefner's models of the human as "eschatological companion" and "created co-creator" respectively will be examined along with Schaefer's "virtuous cooperator"[35] as options for consideration and application to engineers. Most of Borgmann's *Power Failure* will be assigned to immerse students in his thinking about the need for engineers and theologians to collaborate in restraining and redeeming technology and in determining how that need can be filled. An article by Normand Laurendeau will stimulate student thinking about his "ethic of responsibility" for engineers.[36] This section will end with the identification of theological foundations and norms for human agency that are needed in subsequent sections. While the theology professor will take the lead, the input of the engineering professor is vital to question, inform, and challenge theological reflection and to develop the model of human agency to assure they are pertinent to the engineering profession. Each student will opt for one model and explain why in a reflective essay.

The third section of the seminar will be dedicated to examining one engineering project in sufficient depth and reflecting theologically on that project from the primarily Catholic foundations and ethics identified in the prior section. Of course, the engineering professor will take the lead on explaining the project, while the theologian will become sufficiently informed about it so theological reflection can be relevant, meaningful, and useful when addressing the project from a theological and ethical perspective. This section will serve as the exemplar for the culminating section of the seminar.

For the last section, students will be teamed in pairs to research and represent the engineering or theological perspective on an engineering project they have selected from among possibilities identified early in the seminar. Students will choose the religious tradition upon which they wish to reflect—an option that facilitates investigating their own traditions. Their research and planning for this task will have begun before the third section of the seminar commences, and consultation sessions with the pertinent professor will be mandatory to assure that the students are on the right track and working well with one another as mutually engaged disciplinarians. The engineers and theologians will present their findings in seminar, and the other students will provide helpful critiques of the adequacy of the presenters' representations of their assigned disciplines. As a culminating project, each student will write a reflection paper in which the engineering and theological perspectives on the project are integrated.

Reasons for interfacing theology and engineering are compelling during our age of technology, and particularly so at Catholic universities that strive to provide a unifying educational experience for their students. A graduate seminar on theology, technology, and ethics at Marquette University offered an opportunity to test the fruitfulness of dialogue. Lessons learned from this seminar will be applied to a future seminar in which theology and engineering will be interfaced by a team of theology and engineering professors who will demonstrate how their disciplines relate constructively to one another. This future seminar will continue to present major challenges to teaching, especially in light of the paucity of scholarly literature on the history of the relationship between theology and technology in general and engineering in particular. Established theologians who are seeking fruitful avenues of research and students who are engaged in theology doctoral programs should be stimulated by the need and many possibilities for pursuing the interface of theology, technology, and ethics.

NOTES

1. Core of Common Studies [Statement], Marquette University, available at http://www.marquette.edu/programs/core.

2. Among the examples of providing interdisciplinary experiences are Origin and Nature of the Universe, a course taught jointly by a physicist and a systematic theologian specializing in the theology–natural sciences relationship, and the capstone seminar for the Interdisciplinary Minor in Environmental Ethics, a course that draws upon the knowledge and skills learned in specific biology, economics, philosophy, physics, environmental engineering, and theology courses to address an environmental problem.

3. Pontifical Biblical Commission, *Interpretation of the Bible in the Church* (Washington DC: United States Catholic Conference of Bishops, 1993).

4. Ian Barbour, *Ethics in an Age of Technology,* Gifford Lectures 1990–91 (San Francisco: HarperSanFrancisco, 1993).

5. These principles of Catholic social teaching have been identified by the *magisterium* of the Roman Catholic Church and reflected upon by scholars, especially since the encyclical by Pope Leo XIII, *Rerum novarum.* For the central social teachings by pontiffs and councils, see the Pontifical Council for Justice and Peace, "The Social Agenda" (http://www.thesocialagenda.org/) and *Compendium of the Social Doctrine of the Church* (Washington DC: United States Conference of Catholic Bishops, 2004). For scholarly reflections, see Edward P. DeBerri and James E. Hug, *Catholic Social Teaching: Our Best Kept Secret,* 4th ed. (New York: Orbis Books, 2003). Among the most erudite sources on the often reflected upon notion of the common good are the following: David Hollenbach, *The Common Good and Christian Ethics* (New York: Cambridge University Press, 2002); James Donahue and M. Theresa, eds., *Religion, Ethics, and the Common Good* (Mystic, CT: Twenty-Third Publications, 1996);

David A. Boileau, ed., *Principles of Catholic Social Teaching* (Milwaukee: Marquette University Press, 1998); Dennis P. McCann and Patrick D. Miller, eds., *In Search of the Common Good* (New York: T & T Clark, 2005); Timothy Backous and William C. Graham, *Common Good, Uncommon Questions: A Primer in Moral Theology* (Collegeville, MN: Liturgical Press, 1997). On the notion of valuing the physical creation, see Jame Schaefer, "Valuing Earth Intrinsically and Instrumentally: A Theological Framework for Environmental Ethics," *Theological Studies* 66, no. 4 (2005): 783–14.

6. Jame Schaefer, "The Virtuous Cooperator: Modeling the Human in an Age of Ecological Degradation," *Worldviews: Environment, Culture, Religion* 7, nos. 1–2 (2003): 171–95.

7. For Aquinas's treatment of the virtue of prudence, see, for example, *Summa theologiae* 1I–II.65 and II–II.47.

8. On temperance, see Aquinas's *Summa theologiae* II–II.141.

9. On justice, see Aquinas's *Summa theologiae* II–II.58 and 61.

10. On fortitude, see Aquinas's *Summa theologiae* II–II.123.

11. Thomas Aquinas, *Summa theologiae* II–II.23-25; also *De caritate* 3 and 7.

12. Pope John Paul II, *Fides et ratio, an Encyclical Letter to the Bishops of the Catholic Church on the Relationship between Faith and Reason,* no. 106, Vatican City, September 14, 1998.

13. See the books of Proverbs, Ecclesiastes, Song of Songs, Ecclesiasticus, Sirach, and Wisdom.

14. Barbour, *Ethics in an Age of Technology*, 3–25.

15. See Pope Leo XIII, *Rerum novarum, Encyclical on the Condition of the Working Classes,* Vatican City, May 15, 1891, and Pope John Paul II, *Laborem exercens, Encyclical on the Ninetieth Anniversary of Rerum Novarum,* Vatican City, September 14, 1981.

16. Pontifical Biblical Commission, *Interpretation of the Bible in the Church,* 40–42; see also 72–75 for a strong critique of the fundamentalist approach whereby the words are taken literally instead of attempting to understand them in the context of the times and understanding of the world in which they were written.

17. Karl Schmitz-Moormann in collaboration with James F. Salmon explores poignantly the concept of *creatio appellata* in their *Theology of Creation in an Evolutionary World* (Cleveland: Pilgrim Press, 1997).

18. Examples of traditional topics include ecclesiology, pneumatology, and church and state; examples of specialized topics include St. Augustine's *De Trinitatae, nouvelle theologie,* and eucharistic controversies.

19. David Noble, *The Religion of Technology: The Divinity of Man and the Spirit of Invention* (New York: Alfred A. Knopf, 1997); Susan White, *Christian*

Worship and Technological Change (Nashville: Abingdon Press, 1994); Jay Newman, *Religion and Technology: A Study in the Philosophy of Culture* (Westport, CT: Praeger Press, 1997); David Hopper, *Technology, Theology, and the Idea of Progress* (Louisville: Westminster/John Knox Press, 1991).

20. Noble, *The Religion of Technology*, 10.

21. Noble, *The Religion of Technology*, 206–7.

22. Susan White's book provides additional support for this dialogue, although it implies that a background in the history of technology is required in order to understand fully the practices of the church. See further her discussion of calendars and clocks in chapter 4, "Worship and Technology in History, 1: Medieval Engineering and the Liturgy," in *Christian Worship and Technological Change*.

23. Willem Drees, "Human Meaning in a Technological Culture," and "'Playing God? Yes!' Religion in the Light of Technology," *Zygon* 37, no. 3 (2002): 597–604 and 643–54; Carl Mitcham, "Technology as a Theological Problem in the Christian Tradition," in *Theology and Technology: Essays in Christian Analysis and Exegesis*, ed. Carl Mitcham and Jim Grote, 3–20 (Lanham, MD: University Press of America, 1984); Jacques Ellul, "The Technological Order," in *The Technological Order: Proceedings of the Encyclopedia Britannica Conference*, ed. Carl F. Stover, 10–37 (Detroit: Wayne State University Press, 1963), and "Technology and the Gospel," *International Review of Mission* 66, no. 262 (1977): 109–17; Robert John Russell, "Five Attitudes Toward Nature and Technology from a Christian Perspective," *Theology and Science* 1, no. 2 (2003): 149–59; Philip Hefner, *Technology and Human Becoming* (Minneapolis: Fortress Press, 2003); Paul Tillich, *The Spiritual Situation in Our Technological Society*, ed. J. Mark Thomas (Macon, GA: Mercer University Press, 1988).

24. For those (like ourselves) who are interested in pursuing dialogue between theologians and engineers, there is much to be learned from the past and present experiences of theologians and scientists engaged in dialogue. An examination of the similarities and differences of issues at the boundaries of these disciplines is beyond the scope of this essay.

25. Ellul, "Technology and the Gospel," 117.

26. Hefner, *Technology and Human Becoming*, 87.

27. In "Five Attitudes Toward Nature and Technology from a Christian Perspective," 157, Russell shares his model of the human as an "eschatological companion" in which he incorporates Hefner's "created co-creator," thereby recognizing the human ability to alter nature and humanity through technology while acknowledging that this power has limits since God, not humankind, will usher in the eschaton.

28. Hans Jonas, *The Imperative of Responsibility: In Search of an Ethics for the Technological Age* (Chicago: The University of Chicago Press, 1984); Albert

Borgmann, *Power Failure: Christianity in the Culture of Technology* (Grand Rapids, MI: Brazos Press, 2003); Murray Jardine, *The Making and Unmaking of Technological Society: How Christianity Can Save Modernity from Itself* (Grand Rapids, MI: Brazos Press, 2004).

29. Jardine, *The Making and Unmaking of Technological Society*, 9.

30. A model for this approach is the team-taught course at Marquette University, Origin and Nature of the Universe, which has been offered four times over the past eight years and which juniors and seniors could take for physics or theology credit.

31. A survey of publishers is continuing to determine the future availability of the kind of text we envision.

32. John M. Staudenmaier, *Technology's Storytellers: Reweaving the Human Fabric* (Cambridge MA: MIT Press, 1985); John Brooke and Geoffrey Cantor, *Reconstructing Nature: The Engagement of Science and Religion*, Gifford Lectures 1995–96 (Edinburgh: T & T Clark, 1998).

33. John Staudenmaier, S.J., "Recent Trends in the History of Technology," *The American Historical Review* 95, no. 3 (1990): 715–25.

34. Philip Hefner, "Technology and Human Becoming," *Zygon* 37, no. 2 (2002): 655–66.

35. Schaefer, "The Virtuous Cooperator."

36. Normand Laurendeau, "An Ethic of Responsibility for the Engineering Profession: Philosophical and Religious Perspectives," *CTNS Bulletin* 20 (2000): 3–20.

PART 3

International Service Learning

Chapter Seven

Human Development and a Senior Project in Mali

CAMILLE M. GEORGE AND

BARBARA K. SAIN

The University of St. Thomas (UST) has two innovative capstone courses available for engineering students. One course is the Engineering Design Clinic, the senior capstone component of the engineering curriculum. Each year, one or two senior design teams are engaged in global service-learning projects that seek to address a specific engineering need in the developing world. Recently one of those projects was to develop a passive cooling system for use in Mali and other areas of sub-Saharan Africa. The second course is an upper level theology course entitled Christian Faith and the Engineering Profession that explores the relation of theology and engineering, particularly how the two disciplines come together in the profession of engineering lived in the contemporary world. Study of theological topics, such as the dignity and integral development of the human person, provides a foundational Christian vision for consideration of ethical issues in engineering. Extensive case study analysis assists the integrated comprehension of that vision and its application to the social, moral, and professional obligations and responsibilities of engineers.

This chapter draws on both courses, using human development as a unifying theme. A theological understanding of development, drawn from Catholic social teaching and presented in the theology course, will be used in this paper to analyze the experience of UST faculty and students involved in the development project in Mali. The senior design project in Mali is used here as an example of the case study analysis that takes place in the theology course. The discussion of theology and engineering in that course gives the students the opportunity to integrate the technical and liberal arts components of their education.

The remainder of this introduction describes the two capstone courses. The following section contains the case study analysis. It discusses the relevant theological principles, describes the cooling project in Mali, and concludes with an analysis of the project. A final section places the courses and the representative case study in the context of Catholic education, highlighting the contribution Catholic universities can make to the integrated education of the students in their engineering programs.

Integration of Global Service Learning in the Senior Design Capstone

The senior design course is an integral part of the undergraduate engineering curriculum. The School of Engineering at UST has placed a large emphasis on the practical aspects of the discipline. As a Catholic university it has the opportunity to provide unique design experiences that incorporate service to the community and enable the program to live out its mission as a university concerned with social justice for all. The School of Engineering offers one or two global service-learning projects yearly. These projects seek to address a specific engineering need in the developing world. The students work on the project for two semesters. The first semester culminates in a critical design review of a proposed solution, and the second semester concentrates on the prototype build and test. A final design and comprehensive report is required at the completion of the course sequence. For the past two years the overseas on-site travel has taken place during spring break. The spring-imbedded trip is necessary for prototype evaluation and testing. These projects emphasize developing sustainable practices and introduce the students to global issues. The students must consider the perspectives of another culture as well as the environmental and social impact of their design. The projects require the students to play an active role in the world around them.

The on-site interaction with local communities creates powerful lifelong experiences that can move students to think of their career as a calling rather than just a way to earn a living. The students are asked to think of their talent to grasp scientific concepts and to think analytically as a gift that can be used to better someone's life. During the trip, students are expected to write in a daily trip log. Reflective questions are provided to help the students discern their experiences. These trip-logs have been used in a research study that examines the effect of short-term international service learning on the growth of intercultural awareness or intercultural competency.[1]

Integration of Theology and Engineering in an
Upper-Level Theology Class

UST offers a series of upper level theology courses on Christian faith and specific professions. Christian Faith and the Engineering Profession is the sixth course in this series. One of the objectives of these courses is to facilitate the integration of the university's core curriculum with the specialized coursework of particular majors, especially in the university's professional schools. The recent introduction of a course for engineering majors coincides with an increased emphasis on ethical education in the professional engineering community. The requirement under the Accreditation Board for Engineering and Technology's engineering criteria (ABET EC 2000) to demonstrate clearly that graduates of an accredited engineering program have "the broad education necessary to understand the impact of engineering solutions in a global and societal context" (3.h) and "an understanding of professional and ethical responsibility" (3.f) has resulted in requiring all accredited university engineering programs to have an ethics component in their curriculum.[2] This emphasis on ethics is the result of growing awareness within the professional engineering community that engineers face ethical as well as technical challenges.

Christian Faith and the Engineering Profession discusses a range of engineering issues in the context of the Christian theological tradition. Students are asked to consider the significance of engineering as a profession and the qualities that are required for a "good engineer." From the personal perspective, this discussion allows students to reflect on engineering as part of a vocation in which various aspects of their lives can be integrated into a coherent whole. In a broader perspective, the professional character of engineering brings with it the privileges and responsibilities of ever-greater influence in the global arena. Drawing on the Christian theological tradition, and particularly on Catholic social teaching, the course presents a vision of engineering as part of an integrated life in global society. Extensive case study analysis familiarizes the students with issues in modern engineering and gives them the opportunity to apply the theological ideas they have learned.

Both courses, the senior design projects for the developing world and Christian Faith and the Engineering Profession, offer unique opportunities for engineering majors to deepen their awareness of engineering as a calling that can be integrated with a life of faith. The senior design projects offer practical engagement, service learning, and insight into development efforts in the economic and political arenas. The theology course presents ideas from the tradition of Catholic theology and allows the students to apply those ideas to situations faced by engineers.

THE DEVELOPMENT PROJECT IN MALI: A CASE STUDY ANALYSIS

In the following analysis, some ideas from the course Christian Faith and the Engineering Profession are applied to the Mali development project. The discussion is not a comprehensive or final analysis of the ongoing project and its complex context. Rather, it is intended to demonstrate how students are encouraged to apply the theological principles they have studied to concrete engineering projects. In the course, this discussion occurs in the context of the broader discussion of Christian theology and the engineering profession.

Development in Theology

Many Christian theologians consider development an important theological topic and believe that they can contribute to the broader discussion of development in today's world. The foundation for Christian theological reflection on development is made clear in *Justice in the World,* a statement from the 1971 worldwide synod of Catholic bishops: "Action on behalf of justice and participation in the transformation of the world appear to us as a constitutive dimension of the preaching of the Gospel, or, in other words, of the Church's mission for the redemption of the human race and its liberation from every oppressive situation."[3] Being concerned about justice, peace, and the full development of human beings is not a secondary topic in Christianity. It is at the very heart of the faith. Every Christian should be involved, in ways appropriate to his or her situation, in the ongoing struggle to create a world where all people can flourish.

While Catholic bishops and other Christian leaders proclaim development to be a central concern for Christians, they also acknowledge that, as theologians, they do not have definitive expertise in economics, politics, or the other disciplines helpful for designing effective development policies. The role of Christian theologians is to provide a vision to guide the formulation of policies or the evaluation of existing policies: a vision of human dignity rooted in the gospel. "The Church does not have *technical solutions* to offer for the problems of underdevelopment. . . . But the Church is an 'expert in humanity.'"[4] In this chapter, we draw from some of the classic documents of Catholic social teaching to evaluate the development initiative in Mali.[5]

When Catholic social teaching addresses the specific topic of development, the emphasis is on an integrated development that fosters all aspects of the human person, in the context of the larger community. "Modern underdevelopment is not only economic, but also cultural, political, and simply human."[6] If underdevelopment affects many dimensions of the human person, then any solution must have a similarly comprehensive vision. In 1967 the encyclical *Populorum progressio* ("On the Development of Peoples") introduced the term "integral development" to describe the kind of comprehensive development that is needed. Integral development is development of an individual's entire potential, including the intellectual, cultural, social, emotional, and physical dimensions of his or her existence. This understanding of human development is based on another concept from the Christian theological tradition: human dignity.[7] Each person is created in the image of God and is called to a unique relationship with God and others. Human dignity can be threatened not only by poverty and disease, but also by the loss of autonomy, freedom, and justice.

One way to foster true human development is to encourage people in developing countries to be agents of their own development: to contribute their personal talents and the unique insights of their culture to development planning so that the resulting programs will be effective in their situation. This principle does not preclude substantial assistance from the international community. The point is that active cooperation between development agencies and their intended beneficiaries can foster the human dignity of those in need of assistance and increase the practical effectiveness of the programs. There cannot be a universal blueprint for development; rather general ideas should be combined with the experience of indigenous people to design solutions that fit the local situation. In 1963, Pope John XXIII stated, "The countries being developed . . . are primarily responsible. . . . [They should be] the principal artisans in their own economic development and social progress."[8] People from other nations who desire to assist with development should view their role as one of partnership, working together with those who suffer from underdevelopment in ways that ultimately lead to their independence from any assistance.

The term solidarity has been used in Catholic social teaching to describe relationships of partnership and respect that foster human dignity and contribute to the common good.[9] In 2001 the Catholic bishops of the United States issued a document entitled "A Call for Solidarity with Africa." In this statement, the bishops draw special attention to the situation of Africa, highlighting both the richness of African contributions to world culture and the desperate poverty that afflicts millions on the continent. They emphasize the responsibility of American Catholics to act decisively, in solidarity with African Catholics, to promote integral development in Africa. "We encourage the Catholic community in the United States to contribute its diverse talents and gifts to the continent's causes of justice, peace, and integral development."[10] In order to bring about integral development, the relationship between the first world and the third world must go beyond the simple transfer of money and goods. A 1987 papal encyclical, *On Social Concern,* describes solidarity as a virtue rooted in awareness of the moral implications of interdependence:

> It is already possible to point to the *positive* and *moral value* of a growing awareness of *interdependence* among individuals and nations. The fact that men and women in various parts of the world feel personally affected by the injustices and violations of human rights committed in distant countries, countries which perhaps they will never visit, is a further sign of a reality transformed into *awareness,* thus acquiring a *moral* connotation. . . . When interdependence becomes recognized in this way, the correlative response as a moral and social attitude, as a "virtue," is *solidarity.* This then is not a feeling of vague compassion or shallow distress at the misfortunes of so many people, both near and far. On the contrary, it is *a firm and persevering determination* to commit oneself to the *common good;* that is to say to the good of all.[11]

Interdependence, mutual commitment, and perseverance in working for the common good are all elements of solidarity. In various ways, Catholic leaders have urged all people of good will to work together in solidarity for peaceful, just solutions that foster the full and unique development of all people.

It has become increasingly clear in recent decades that development programs must also respect the natural environment. In 1987, Pope John Paul II argued that "a true concept of development cannot ignore the use of the elements of nature, the renewability of resources and the consequences of haphazard industrialization—the three considerations which alert our consciences to the *moral dimension* of development."[12] Theologically, the moral dimension of environmental issues is rooted in the inherent goodness of creation and human responsibility to care for it. A growing number of Christian theologians are articulating a gospel vision of the stewardship of creation.[13] Development programs intended to promote long-term change should include the use of natural resources as a significant factor in their decisions. Products or programs that are effective in one part of the world may not be equally effective in another, especially when there is a difference of climate or natural resources. Some approaches may be simply unacceptable, even if they meet a short-term need, because of their long-term effect on the environment.

Before applying these principles to the development project in Mali, it will be helpful to discuss that project.

The Development Project in Mali

Agencies and Program Goals

Large sums of money are spent annually by governmental agencies on development projects intended to promote human development in today's world. Recently the US government, through the United States Agency for International Development (USAID), provided over $38 million of assistance to five developmental aid agreements to Mali in West Africa.[14] The sub-Saharan nation of Mali is one of the poorest nations in the world. In 2003, Mali ranked 172 out of 175 countries in the United Nations Human Development Index.[15] Its human development index has been dropping, and social well-being indicators such as adult literacy and children attending school are among the worst in the world. The difficulties faced by the people of Mali are worsened by the effects of drought. Today Northern Mali, Niger, Chad are all experiencing extreme conditions. Millions of people are at risk of starvation.

Included in the USAID agreements for assistance to Mali was a five-year program to hasten development by making information accessible. The "Communications for Development Special Objective" outlined in their "Country Strategic Plan" was as follows:

> Objective: Accelerate development by making information accessible through innovative communication techniques and appropriate tools.[16]

The challenges to this USAID objective include a scarcity of libraries, a low literacy rate, and a limited amount of information on development topics in local languages. The USAID strategy in accomplishing their objective focuses on the convergence of high and low technology by introducing digital technologies, the Internet, and rural radio.

As a concrete step in achieving their goal, USAID funded Project CLIC and built thirteen Community Learning and Information Centers, or Centres Locaux d'Information et de Communication (CLICs) across Mali. These twenty-first century computer-based libraries or "public access tele-centers" were meant to provide useful visual information with topics ranging from health and agriculture to civics and business. In addition, they are to offer basic adult literacy education.

Prior to 2005, USAID/Mali government supported financially the CLICs. Since then, the CLICs have been sold to and run by local independent Malian organizations. As part of their strategy, the USAID/Mali communication development team leveraged additional resources from private industry, non-governmental organizations (NGOs) and the government of Mali in various public-private partnerships and alliances. The core partners in the implementation of Project CLIC were the Academy for Educational Development (AED), based in Washington DC, and the Institut Africain de Gestion et de Formation (INAGEF), an independent Malian NGO. Technical content and expertise for the CLICs was solicited from a variety of organizations, such as World Links, the Peace Corps, the *geekcorps* (a Canadian Informational Technology NGO), and various American universities. The local hosts were "selected based on their capacity to be available to the entire community without discrimination, and also their ability to effectively manage the CLIC." Self-sustainability was a critical program goal; many of the CLICs, however, have since closed.

The CLIC Cooling Project
The University of St. Thomas became involved in Project CLIC through a United States Department of Agriculture Cooperative State Research, Education, and Extension Service (USDA-CSREES) Higher Education Challenge Program grant entitled "Discovery-Based Undergraduate Opportunities: Facilitating Farmer-to-Farmer Teaching/Learning" awarded to Montana State University. The three-year grant funded by the USDA aimed to provide research opportunities to undergraduates as well as

encourage careers in agriculture. The grant involved four universities in a partnership that provided "information content through innovative techniques" for the CLICs by engaging university students through senior-level coursework and independent study. The School of Engineering at UST was subcontracted to examine alternative cooling strategies for the CLICs through its senior design capstone course.

The CLICs are one-room buildings that house four to eight computers and other audio-visual equipment such as printers, DVD players, cameras, and copiers. Technical equipment, humans, and solar radiation all add to the heat load of the building. Project CLIC purchased commercially available evaporative coolers (swamp coolers) as their climate control device. UST was asked to search for alternative cooling options or modify the chosen evaporative coolers. Throughout the developing world there are few solutions to cooling spaces with high internal heat loads. Very little has been published in this area. A search of the literature yielded one similar project involving a low-cost alternative cooling system for a computer building in El Salvador. A team of students from Penn State used a fin-tube liquid-to-air heat exchanger with ground water as the working fluid.[17] Most of the CLICs in Mali are not located near readily accessible ground water, and thus this option was rejected.

Our culture takes for granted comfortable and conditioned public spaces. Unfortunately, our society's solution for cooling space is power intensive, inefficient, based on toxic substances (refrigerants), and untenable for places like sub-Saharan Africa. Electricity is both expensive and unreliable throughout Mali. It is not practical to simply install air-conditioning units.

Technical Challenges of the CLIC Cooling Project

A team of four senior engineering students worked on the design of a low energy cooling system for the Malian CLICs. They were asked to design a culturally appropriate solution that could potentially benefit communities all over Mali and other sub-Saharan countries. The project was approached from three perspectives: modifying the existing evaporative cooler to decrease its power and water consumption; connecting the evaporative cooler to a geothermal duct that could pre-cool the incoming air; and adding strategic shading to minimize the solar heat load on the building. All three approaches were experimentally verified on location in Mali and in the engineering laboratories.

A 20 meter duct was buried at a depth of 0.5 meters before our arrival in Mali. Actual ground measurements and inlet and outlet air temperatures were measured over a three-day period. Results indicated a 10°F drop in incoming air to the evaporative cooler at the hottest time of the day. Details of the experimental study have been published.[18] A solar reflective fabric that prevents more than 90 percent of the sun's radiation from reaching the building and subsequently decreases the building's total interior heat load was installed on the roof of the test-site CLIC. Roof and room temperature measurements, air humidity, and air velocities were monitored over a five-day period, examining both the building baseline and building with intervention. Results showed a decrease in roof temperature of 70°F, significantly altering comfort levels inside the building. A final design recommendation that should decrease room temperatures by 20°F from ambient, use half the electrical power, keep the relative humidity under 50 percent, and consume only one gallon per hour (GPH) of water is documented in the student final report and can be at obtained at the project website.[19] Table 7.1 provides an overview of the technical solution. The shading strategy was ultimately not recommended for most of the CLIC buildings.

Table 7.1 Technical Design Achievements

Engineering Specification	Original Evaporative Cooler	Modified System (Duct + Cooler)
Water usage, gallons per hour	3	1
Power consumption, watts	303	146
Total temperature decrease from ambient due to system intervention (inlet to room)	$\Delta T = 15°F$	$\Delta T = 30°F$
Air flow, cubic feet per minute	105	240

Evaporative Cooling

Although UST was contracted to provide designs that included evaporative cooling, questions can be raised about the ethics and sustainability of a design using water, a precious resource in the drought-prone and dusty region of sub-Saharan Africa. The principal behind evaporative cooling is to pass fresh, dry outside air through a water soaked pad. The energy needed to evaporate the water is taken directly from the air. As a result cooler but more humid air exits the device. Water is usually dripped continually through a porous pad. Power is needed to run the fan and a water pump. These devices work very well in hot and dry climates and, although they cannot achieve refrigeration temperatures, they can cool down rooms to human comfort levels and temperature-humidity zones acceptable for electronic equipment.[20]

The window unit Alpine® Model RW3000 evaporative coolers purchased by Project CLIC were made in Mexico. In the US the model costs about $350.00 plus shipping. They have pads made from aspen, a fast growing tree that is common and inexpensive in North America but does not grow in Mali. No provisions were made by Project CLIC to offer a locally available pad substitute. Thus, as supplied, the Malian CLICs will need to indefinitely import the natural fiber porous pads. Water quality is important to the longevity and performance of any evaporative cooler. The pad replacement schedule is directly related to the quality of water. In the American southwest, evaporative coolers are not installed without appropriate water treatment systems. Coolers are supplied with water of at least the same quality as drinking water.[21] It is recommended to replace the natural fiber pads every year if the cooler is used frequently to avoid potentially harmful biological growth in the pads.

It is unlikely that "drinking quality" water will be used in the Malian CLICs. Clean water is used sparingly throughout Mali. In Bamako, the capital city, tap water is not of drinking quality. Only bottled water would be at the same quality as water used in the American southwest. As built, the evaporative coolers use three gallons of water per hour. Clean bottled water costs about $0.50 per liter. It would be prohibitively expensive and wasteful to use bottled water to cool a computer room. There is no available evidence that Project CLIC has considered the pad life expectancy. The pad replacement regime will vary greatly depending if tap water, well water, or river water is used. It is uncertain if Project CLIC has even established or recommended an appropriate maintenance procedure. It is also doubtful that the current CLIC managers have been properly trained to monitor the aspen pads for biological contamination or are even informed of the potential health risks of using a tainted pad. This is especially troublesome considering the CLICs are to be sold as self-sustaining enterprises.

The pads also act as air filters trapping particles on their wet surfaces. Installation of a blow-down system or a bleed-off kit, to minimize scale build-up, is recommended by the manufacturer in especially dusty areas.[22] At the edge of the Sahara desert, Mali is hot, dry, and dusty for most of the year. Blow-down systems have not been purchased for the CLICs. It is unknown how quickly the pads will become clogged. Without a blow-down system the evaporative coolers will not function effectively, and, consequently, it will be necessary to replace the imported aspen pads more often. Without a maintenance plan, the evaporative coolers may even cease to work. Pads may need to be replaced multiple times a season.[23] Many projects in the developing world have failed because of the lack of "spare parts." Though evaporative coolers are technologically simple and easy to use, there is a minimum amount of maintenance (pad replacement) and caution (pad contamination) required, as well as other long-term issues, such as rust and wiring (the wiring sits in a very humid environment) that must be communicated to the lay user.

Communication Challenges of the CLIC Cooling Project

Mali is a French speaking country. The UST team included a French professor and two senior level language students. Their help on-site was commendable. However, it should be mentioned that the language barriers, though not insurmountable, made all communication difficult.

In the beginning of the project it was believed that the American and French engineering students would be in contact via e-mail with engineering students at the Institute of Engineering, l'Ecole Nationale d'Ingenieurs (ENI), at the University of Bamako in Mali. It was hoped that our Malian counterparts would help in executing the testing phase and become involved with choosing native materials for a locally made evaporative cooler.

The UST principal investigator had introduced herself to the Dean of Engineering and a group of engineering faculty the previous year. The Malian professors were warm and expressed a sincere desire to collaborate on potential projects. Despite several e-mails sent during the fall semester describing the project and soliciting collaboration, there was no response from anyone at ENI before our arrival. The university had only had the internet for one year, and it is highly possible that some of the professors were not entirely comfortable with electronic communication. Mali has a rich history in oral communication through music and story-telling. After eleven years of experience in working in Mali, another researcher from the USDA grant communicates only by telephone.[24] Unfortunately, the need for a translator, and hence three-way calling, further complicates the overseas communication.

Knowing that Mali is a culture more accustomed to face-to-face meetings, the UST principal investigator again contacted ENI's two Malian thermodynamics professors after arriving in Mali. One professor had previously experimented with local substitute materials for evaporative cooler pads and was particularly interested in the project. The UST team invited the professors and their students to join the CLIC cooling testing. However, our on-site travel (spring break) occurred the week before Easter. Even though Mali is more than 95 percent Muslim, the university calendar is based on the French (historically Catholic) system. Consequently, the Institute of Engineering was closed during our entire stay in Mali, the Easter holy week. In fact, there were no students at all on campus. Secondarily, the INAGEF director arranged for us to conduct our on-site tests at a CLIC three hours outside the capital accessible only by four-wheel drive. The professors did not have access to or the funds to rent a four-wheel drive. As a result they were unable to travel to the test site. Interestingly, on the last day of our trip, a Canadian information technology specialist from the *geekcorps* took us to a typical CLIC building only thirty minutes from the capital on a major paved road. It is certainly a location that both faculty (by car) and students (by bus) from ENI could have reached.

Having many institutions involved in this project further complicated communication. For example, the students needed to obtain basic information such as the dimensions of an average CLIC to size their proposed geothermal duct. The institutions with a vested financial interest in our project, the USDA and Montana State University, had no knowledge of the technical specifications of our particular project. USAID/Mali and AED had a peripheral technical interest in our project but also had a limited staff to answer the specific details involved with student e-mails. INAGEF, the Malian organization charged with the day-to-day running of the CLICs, was very helpful in providing information, but as it turned out lacked a fundamental understanding of our role in this rather complex web of development providers.

Even though the ex–Peace Corps volunteer who had been assigned to work with INAGEF headquarters the previous year was our project liaison in Minnesota, there was a fair amount of miscommunication. UST was working under the assumption that they were to ultimately provide a design recommendation for a typical CLIC. On-site experimental data was to be used to verify their proof of concept. INAGEF gave us permission to test our concepts at one of the CLICs and facilitated the digging and laying of a geothermal duct before our arrival. The CLIC they chose, to our surprise, was not a typical CLIC, but their largest CLIC and the one most in need of a cooling strategy. The CLIC located in Kangaba (three hours from the capital) was not a small one room building, but a large multi-room building in which one of the rooms housed the computer center. The dimensions of the CLIC given through the internet were the interior dimensions of the room. The roof shading strategy designed by the students needed to be completely reengineered on-site. There were numerous surprises not anticipated by the team. The roof had large protruding bolts which easily ripped the solar reflective fabric. The building had a drop ceiling, which required a complete alteration of the numerical analysis. Both the experimental set-up and data gathering needed creative adjustments.

Even more problematic was the fact that the building was built before the arrival of the evaporative cooler. To prevent theft, bars have been permanently installed on the windows. Thus, the evaporative cooler was not and could not be installed. The geothermal duct to evaporative cooler could not be tested as planned. The UST team needed to improvise, and the evaporative cooler was only used as a fan to pull air through the geothermal duct and obtain ground pre-cool measurements. Actual cooling of the computer room was ultimately not possible.

In both of the CLICs visited by the UST team, the purchased evaporative coolers were sitting unused in a storeroom. One of the CLIC employees had tried to run the cooler *inside* the room but he told us that the room became more and more uncomfortable. If you re-humidify the same air, you eventually have saturated or 100% relative humidity, a most uncomfortable situation. The employee did not understand how an evaporative cooler worked and had no "training" on its use or maintenance.

A follow up with INAGEF revealed that the hosts at the Kangaba CLIC were disappointed that UST did not "finish the project." Today, the Kangaba CLIC remains uncooled. The duct lays unconnected, and the evaporative cooler sits in the storeroom. No one has read the final report nor responded to the student installation proposal. The shading barrier left by the student team was ripped apart with the arrival of the first summer rains.

The UST team returned in 2007 and connected the duct with the evaporative cooler. It is rarely used because the hosts at the Kangaba CLIC were disappointed in its performance. The shading barrier left by the student team was ripped apart with the arrival of the first summer rains in 2006 and subsequently removed by a Peace Corps volunteer.

Analysis

The previous discussion outlined the technical and communication challenges encountered by UST faculty and students working on the CLIC cooling project. The complexity of the situation makes it difficult to present a nuanced discussion in a short space. Therefore, as was stated earlier, the analysis here is not intended to be comprehensive. Rather, it selects several aspects of the CLIC cooling project and considers them in light of relevant theological principles. The aspects of the projects chosen for analysis are communication, the appropriateness of the project for the intended recipients, the inclusion of indigenous knowledge, and the use of natural resources. The theological ideas are integral human development, solidarity, and stewardship of natural resources.

Communication difficulties were a significant obstacle to the success of the CLIC cooling project. As part of their strategy, the USAID/Mali communication development team purposely leveraged additional resources from various public-private partnerships and alliances. This seems like a good idea, but it made communication difficult. As a small subcontractor distanced by several levels from USAID/Mali, it was difficult for UST to figure out who was responsible for what. The communication difficulties also affected Malians involved in the program. Some representatives of local agencies were unaware of UST's role in the project. Others seemed to have an incorrect understanding of what UST would contribute. The USAID/Mali communications team leader and the INAGEF director also seemed unaware that some evaporative coolers delivered to the CLICs were sitting in store rooms. Equipment delivery and storage problems must be resolved by careful design. It is probable that there are too many CLIC subcontractors, and, as a result, no one is responsible for "seeing the big picture." Given also the additional language and cultural communication difficulties, the current complex network of agencies does not seem to be an effective strategy for implementing USAID plans to accelerate development.

The CLICs are to be sold to Malian investors. This is consistent with the USAID goal of establishing self-sustainable programs. It is not clear, however, that learning centers run as for-profit private enterprises will really serve average people. The tele-center library concept seems admirable, but during our stay few people actually used the CLIC. The Kangaba CLIC charged 2000 CFA/hour, which is roughly $4.00/hour, to use the computers. This is more than most people earn for a day of labor. The average people living near the CLIC had little knowledge of its purpose. During our short stay we observed only a small number of well-off patrons checking their e-mail or using the copier. Large sums of money are spent on development, but tele-centers may not be what the Malians need.

Project CLIC may have failed to employ sufficient indigenous talent and knowledge, such as an involvement by the Malian Institute of Engineering. It is uncertain if any Malian engineers at all were contacted by INAGEF about the purchase, installation, or maintenance of the purchased evaporative coolers. Despite the fact that the internet was installed at the Institute of Engineering at the University of Bamako in 2004 with the help of the same USAID/Mali communication development team that supported the CLIC construction, the Institute of Engineering had no prior knowledge of, or involvement in, Project CLIC.

Indigenous knowledge could also have been helpful for the construction of the buildings.[25] Each CLIC was supplied with the same equipment and most were built from a general "easy to build template." Although using concrete blocks and a cement roof simplifies construction at multiple locations, incorporating local "mud hut" materials might have better mitigated solar heat loads. The CLIC building in Ouéléssébougou, Mali, has two windows and both its doors facing southwest. Simply changing the building orientation would have greatly reduced the solar load on this building! Other development projects have successfully applied innovative ideas to similar situations. The best strategy for minimizing building energy usage is to incorporate passive cooling in the original design.[26] "Green" architecture aims to build sustainable structures using both local materials and local conditions.[27] Retrofitting existing structures for passive or low-energy cooling is not an efficient or effective strategy.[28] Some individualized building pre-thought would have been prudent, and knowledge of traditional Malian buildings might have proven a valuable resource.

Finally, the purchase of evaporative coolers to cool the Malian CLICs is not a well-thought-out use of natural resources. On a basic level, cooling a computer room with a precious resource is not good stewardship. Secondly, evaporative coolers are not maintenance free. CLIC managers need to consider whether the coolers will be run with untreated water and will need a specific timetable for monitoring biological contamination. Either the pads must be replaced with imported aspen chips or there must be a plan to assist the Malians in designing an alternative.

A Theological Lens

From the theological perspective, a development initiative can be evaluated based on whether it promotes integral human development. Does the project have a vision of development that respects the dignity of all people involved and includes more than economic considerations? If not, what is missing from its vision? Individual projects need not address all aspects of human development, but each project should at least be compatible with a comprehensive vision of development.

The USAID development initiatives in Mali clearly have economic development as a goal. The CLIC project can be seen as an effort to build on the current agricultural form of the local economy while introducing information and communication techniques typical of the global economy. If local farmers used the CLICs to learn farming techniques, as the development agencies envisioned, the improved yield would benefit local families in the short term and the access to global communication would provide opportunities for long-term benefits. Access to the CLICs could therefore provide educational and social as well as economic benefits. It seems that this project could be compatible with the ideal of fostering integral human development.

Despite the value of its overall goals, the CLIC project has not yet been successful in achieving them. Some insight into what has gone wrong can be gained from considering a criterion often mentioned in theological statements about development: allowing the recipients of aid to be agents of their own development. When people suffering from underdevelopment are included in development planning, their personal development is fostered and they can help to ensure that the resulting programs are practically and culturally effective. The CLIC project in Mali began as part of a large scale international aid program. Although it is clear that the program intended to foster international partnerships and sustainable results, the network of development providers involved became so complicated that it was difficult for individuals to understand how all the others were involved. Communicating effectively was even more difficult. While acknowledging the complexities of the situation, one can still conclude that involving local Malians, particularly INAGEF and the engineers at ENI, could have dramatically improved the planning and implementation of the CLIC cooling project. This would be consistent with the theological principle that people should be agents of their own development.

It was also observed that ordinary people, the intended beneficiaries of the CLICs, were not using them. One difficulty was certainly the cost. Another might be the novelty of computers as a pedagogical tool. If the local Malians are not accustomed to learning with computers, the CLIC project will not be effective without a plan for teaching the people how to use the centers. The knowledge of local representatives, who understand the customs of the people, is vital for planning an effective program. It seems, however, that organizational and communication difficulties have prevented the effective implementation of the CLICs. If the CLICs are used only by the wealthy who are already familiar with computers, the learning centers might broaden rather than reduce economic differences in Malian society. Although the USAID/Mali project objectives state clearly that the CLICs should be available to the entire community, it is not clear whether enough attention was given to how this would be accomplished.

Another criterion mentioned in theological documents is solidarity. Christian leaders have argued that the problems of underdevelopment can only be effectively solved when people are united in ongoing relationships of mutual respect dedicated to the common good. Although the agencies involved in the CLIC cooling project have made some provision for the development of relationships among providers and recipients, the challenges of communication have proven to be a serious obstacle. Even in the midst of these difficulties, it is clear that effective working relationships among engineers in both countries, aid agencies, and local representatives could open the door to dramatic changes in the lives of the local people. Equally important as the practical changes are the possibilities for the intellectual, social, and cultural enrichment of all involved. This potential gives support to the notion that solidarity is indispensable for integral human development.

A final point to be considered is whether the project demonstrates good stewardship of creation, respect for the environment, and effective use of local renewable resources. As discussed above, it seems that insufficient attention was given to the effectiveness of imported evaporative coolers in the dusty sub-Saharan climate of Mali. Consultation with local engineers might lead to alternative cooling methods that are effective and truly sustainable.

What Has Been Learned

In its current form, the CLIC cooling project has clear weakness from the perspectives of both engineering and theology. The broad goals outlined by international development agencies have not been effectively implemented, despite intensive effort by many people involved. It is clear that ineffective communication and organization are a significant part of the problem.

Despite these difficulties, the CLIC project provides the foundation for genuine hope. There is real potential for the financial, intellectual, and personal resources invested in the project to bear lasting fruit for the people of Mali. The first stages of the project, complicated as they have been, have made connections between engineers at UST and representatives of local agencies in Mali, connections that can develop into effective working relationships. By participating in this project, UST professors and students have gained real knowledge of the challenges of engineering for the developing world. They have laid the foundation for partnerships that can lead to genuine solidarity and effective practical solutions. These achievements are consistent with the vision articulated by Catholic social teaching and by the American bishops. They are also in line with the goals of USAID/Mali and other development agencies involved in the project. The project is a first step, both realistic and hopeful, in fulfilling the UST School of Engineering's mission of working for social justice in the world.

THE ROLE OF CATHOLIC UNIVERSITIES

Educational institutions, including Catholic universities, are in an excellent position to promote the integrated education of engineers. To use human development, the theme of this chapter, as an example,

students can be introduced to the topic of development and related issues in a variety of disciplines, including but not limited to economics,
politics, technology, history, anthropology, and business. Engineering
programs have a unique opportunity to contribute because many of
the corporations involved with international development projects
employ engineers. Some of the students who pass through our classes
become the engineers working on projects in the developing world.
"Justice in the World," the 1971 Catholic bishops' statement cited
above, states that education focused on justice can

> inculcate a truly and entirely human way of life in justice, love, and
> simplicity. It will likewise awaken a critical sense, which will lead
> us to reflect on the society in which we live and on its values; it will
> make men ready to renounce those values when they cease to pro-
> mote justice for all men. . . . It is also a practical education: it
> comes through action, participation, and vital contact with the re-
> ality of injustice.[29]

The type of education envisioned here offers students the knowledge
and critical thinking skills to make informed and compassionate decisions. It is education for a calling, rather than simply for a career.

Giving students the opportunity to learn about development
through service learning and classroom discussion provides them with
knowledge of basic facts and allows them to think through complex
issues before they confront them in the workplace. Studying the vision
of integral development outlined in Catholic social teaching helps the
students understand the issues of development, which are discussed in
secular terms in political and economic forums, in the context of the
Christian faith. Without pretending to give expert economic or political advice, the documents of Catholic social teaching offer foundational ideas that can be used to evaluate the merits of specific development proposals. Through Catholic social teaching and the international
service-learning senior design projects, the students can see at least
one vision of how engineering can be used for the benefit of all people.

This awareness contributes to the integrated development of the
students themselves. In *Christian Faith and the Engineering Profes-*

sion, students have the opportunity to combine the knowledge and skills gained in core curriculum courses to the realities of engineering. For many students, it is the first time in their university experience that they have reflected explicitly on the connections between social, political, and spiritual issues and the technical problems they tackle in engineering. Students who are involved in the senior design project and travel overseas gain further awareness through personal experience.

Such awareness can be the foundation of the solidarity for which the American bishops call. The senior design project allows some students to take a first step in developing this type of solidarity: not just a compassion for those less fortunate, which many of them already feel, but a practical response in partnership with those people. The students learn about a local community in Mali, meet some of the people there, and have the experience of working with them. Although brief, the experience initiates relationships and gives the students an understanding of how international solidarity and genuine development can occur. Both courses offered at UST, the senior design projects and Christian Faith and the Engineering Profession, contribute to the integral development of the students and faculty involved, as well as to the progress of human development in the world.

NOTES

We would like to acknowledge funding from the Ireland Fund supported by the Lily Endowment through the Beyond Career to Calling project at UST; support from Montana State University in connection to the USDA-CSREES Higher Education Challenge Program, grant #404-570; and the Exploring Ethics across the Disciplines (EEAD) research initiative at UST. We would also like to thank both Dr. John Abraham and Dr. Ashley Shams for their immeasurable help on-site in Mali.

1. A. Shams and C. George, "Global Competency: An Interdisciplinary Approach," *Academic Exchange Quarterly* 10, no. 4 (2006): 249–56, Winter 2006; available at http://www.rapidintellect.com/AEQweb/cho3581z6.htm.

2. Accreditation Board for Engineering and Technology, *Criteria for Accrediting Engineering Programs* (The Engineering Accreditation Commission of

the Accreditation Board for Engineering and Technology, 2000), available at http://www.abet.org.

3. Synod of Bishops, "Justice in the World," in *Catholic Social Thought: The Documentary Heritage,* ed. David J. O'Brien and Thomas A. Shannon (Maryknoll, NY: Orbis, 1995), 289.

4. John Paul II, "On Social Concern," in *Catholic Social Thought,* ed. O'Brien and Shannon, 41.

5. The phrase "Catholic social teaching" is used to describe a body of statements made by Catholic leaders in the past century that deal explicitly with the implications of Christian faith for broad social issues. Beginning in 1891 with the encyclical *Rerum novarum* ("The Condition of Labor"), Catholic leaders have argued that justice, economic development, working conditions, wages, the political order, and other social issues are topics that Catholics should reflect upon in the light of their faith. Catholics inspired by these documents have found common ground, both intellectually and practically, with Christians of other denominations and with some non-Christians. In recent decades the increasing interdependence of economic, political, and social structures worldwide has made only more apparent the need for reflection upon social issues. The tradition of Catholic social teaching is part of the unique contribution that Catholic universities can make to the education of their engineering students.

6. John Paul II, "On Social Concern," Introduction.

7. Second Vatican Council, "Pastoral Constitution on the Church in the Modern World," in *Catholic Social Thought: The Documentary Heritage,* ed. David J. O'Brien and Thomas A. Shannon (Maryknoll, New York: Orbis, 1995), 11–32.

8. John XXIII, "Peace on Earth," in *Catholic Social Thought,* ed. O'Brien and Shannon, 123.

9. John Paul II, "On Social Concern," 38–40. The use of the term solidarity in Catholic social teaching should be distinguished from its use by particular political organizations, such as the Solidarity movement in Poland in the 1980s.

10. United States Conference of Catholic Bishops, "A Call for Solidarity with Africa" (Washington DC: United States Conference of Catholic Bishops, 2001).

11. John Paul II, "On Social Concern," 38.

12. John Paul II, "On Social Concern," 34.

13. See Drew Christiansen and Walter Grazer, eds., "'And God Saw That It Was Good': Catholic Theology and the Environment" (Washington DC: United States Catholic Conference, 1996).

14. USAid, Sahelian West Africa—Humanitarian Emergency; Fact Sheet #3, Fiscal Year (FY) 2005. August 23, 2005; available at http://www.usaid.gov/

our_work/humanitarian_assistance/disaster_assistance/countries/sahel/fy2005/Sahel_HE_FS_3_8-23-2005.pdf.

15. Human Development Report 2003, available through http://hdr.undp.org/en/reports/global/hdr2003/.

16. "Establishing Community Learning and Information Centers (CLICs) in Underserved Malian Communities, Final Report," available at http://www.eric.ed.gov/PDFS/ED502041.pdf.

17. Thomas H. Colledge, "Alternative HVAC Design of Computer Building, Nueva Esperanza, El Salvador," available at http://www.cede.psu.edu/ed/colledge/computerhvac.html.

18. J. Abraham and C. George, "Micro-Geothermal Devices for Low-Energy Air-Conditioning in Desert Climates," *Geo-Heat Center Quarterly Bulletin* 27, no. 4 (2006): 13–15, available at http://geoheat.oit.edu/bulletin/bull27-4/art5.pdf; and "Full-Building Radiation Shielding for Climate Control in Desert Regions," *International Journal of Sustainable Energy* 26, no. 3 (2007): 167–77.

19. Jennifer Borofka et al., "A Comprehensive, Analytical Study of Low Powered Cooling in Hot Desert Climates, Final Report," available at http://courseweb.stthomas.edu/cmgeorge/mali_cooler/.

20. See http://energyoutlet.com/res/cooling/evap_coolers/; and P. W. Fairey, "Passive Cooling and Human Comfort," Florida Solar Energy Center Publication DN-5, January 1994.

21. See http://energyoutlet.com/res/cooling/evap_coolers/.

22. Adobe Air Window Cooler Specifications, http://www.adobeair.com/info-products-alpine-windowcooler-specs.html.

23. Personal communication with Stan Day, October 31, 2004. He has used an evaporative cooler in New Mexico for twenty years.

24. Personal communication with Dr. Florence Dunkel.

25. Examples of other development projects that addressed similar problems include a project in Tibet using thick walls of adobe and a project in Zimbabwe using rockstores and wind-assisted ventilation. These projects are examples of well-thought-out structures adapted to local conditions. Both employ appropriate technologies based on indigenous materials in a pioneering manner. Other building configurations employing night and natural ventilation have been documented and used extensively in Israel, Libya, Oman, and other hot dry climates. S. Hassid, "Evaluation of Passive Cooling Strategies for Israel," available at http://www.inive.org/members_area/medias/pdf/Inive%5CIBPSA%5CUNOESC568.pdf; P. C. Agrawal, "Review of Passive Systems and Passive Strategies for Natural Heating and Cooling of Buildings in Libya," *International Journal of Energy Research* 16, no. 2 (1992): 101–17; Y. H. Zurigat, H. Al-Hinai, B. A. Jubran, and Y. S. Al-Masoudi, "Energy Efficient Building Strategies for School Buildings in Oman," *International Journal of Energy*

Research 27, no. 3 (2003): 241–53; and N. M. Nahar, P. Sharma, and M. M. Purohit, "Studies on Solar Passive Cooling Techniques for Arid Areas," *Energy Conversion and Management* 40, no. 1 (1999): 89N95.

26. Passive and Low Energy Cooling Survey, available at http://www .buildinggreen.com/features/mr/cooling.cfm.

27. B. Givoni, *Passive and Low Energy Cooling of Buildings* (New York: John Wiley & Sons, 1994); and D. Chiras, *The Solar House: Passive Heating and Cooling* (White River, VT: Chelsea Green Publishing, 2002).

28. N. Bouchlaghem, "Optimizing the Design of Building Envelopes for Thermal Performance," *Automation in Construction* 10, no. 3 (2000): 101–12.

29. Synod of Bishops, "Justice in the World," 296.

Chapter Eight

International Service Learning at Marquette University

DANIEL H. ZITOMER, LARS E. OLSON,
AND JOHN P. BORG

Most engineering curricula focus on science and mathematics to solve problems for groups of people. However, it is clear that knowledge of the group to be served and ethical discernment are also important to plan, design, and construct engineering works successfully. The fact that the social sciences are an integral part of engineering solutions has already been widely recognized. For example, Vesilind wrote about the benefits of reforming engineering education to include more applied social science; virtuous decision making in the workplace is the major outcome described.[1] More broadly, the Accreditation Board for Engineering and Technology (ABET) outlines ten skills that engineering graduates should demonstrate, six of which relate to social science (such as considering engineering solutions in a global and societal context), whereas only four of which relate to engineering and natural science (for example, applying math, science, and engineering knowledge).[2]

To serve society more fully and lead rewarding lives, engineering students must be given opportunities to develop better communication and civic participation skills. William Wulf, former president of the National Academy of Engineering, has described the importance of participation by engineers in public policy debate, having written that engineers must "think about the social and political implications of technology" because engineering has a significant effect on society. Policy decision making will therefore benefit from significant technical input.[3] Wulf goes on to point out that the present engineering culture is indifferent or actively opposed to debate of public policy issues. Wulf believes this situation should change.

Ethical discernment, the perception and judgment of a person's actions in light of standards of conduct, is another skill that employers report as essential for engineers to possess if they are to become leaders.[4] Engineers must not only plan, design, build, and operate systems, they must also consider the ethical consequences of their work, providing leadership in teams and in the community. In his recent book, Chris Lowney applies leadership concepts used for over 450 years by the group of Catholic priests known as the Society of Jesus (the Jesuits). He points out that self-awareness, ingenuity, love, and heroism can be developed through the *Spiritual Exercises* written by the society's founder, St. Ignatius, as well as other reflective techniques. These virtues can serve as foundations for successful decision making.[5]

Self-awareness, in turn, strengthens engineers' resolve to act with "indifference"—remaining open to many methodologies—so as not to be hampered by attachment to one exclusive road that possibly leads to failure. It is important to note that, in the general sense, "indifference" is often assumed to have a negative connotation. People that are indifferent may be thought to be uninspired, uninformed, and disinterested. However, as used by St. Ignatius and the Jesuits, the concept of indifference is inspired, informed, and valuable for attaining the greater good. For example, when designing and constructing civil engineering works, it is often necessary to adapt, make field modifications, and remain "indifferent" so as to best serve a community. On more than one occasion, Marquette University students have spent three months designing a bridge, only to travel to rural Guatemala and find that the community members prefer a much higher bridge deck to accommodate greater river flows. After much debate, a compromise is typically reached. The bridge deck may not be raised from five to twelve feet, but raising it to eight feet is reasonable, and the bridge can be constructed. The design must then be quickly modified. If the student designer and supervising professional engineer remain "indifferent," then they find that the redesign is not a reason to panic, but the compromise is for the best both technically and socially.

In most cases, pedagogical methods linking social science, ethical discernment, and engineering are underdeveloped. Difficulties have been described as disinterest and cynicism of some educators towards engineering ethics and social science initiatives, a lack of engineering faculty who are committed to including ethics and social sciences in class, and a lack of motivation among students to learn about these subjects.

SERVICE LEARNING AS A WINDOW OF OPPORTUNITY IN ENGINEERING EDUCATION

Because engineers stereotypically resist debate about public policy as well as reflection on their work's social dimensions, specific methods are required to engage engineering students in these broader issues. One emerging method involves service learning and the related pedagogy of community-based research, described below. Engineering students enthusiastically grasp hands-on projects in which a final product is designed and built. If the hands-on project is accomplished in association with a community that requests the work, then the experience can be the window through which social dimensions enter the classroom. In this way, a service-learning project is an experience that engages engineering students in social as well as technical issues. For example, many Marquette civil and environmental engineering students have been immediately intrigued by the thought of designing and building a bridge in Guatemala. Once they travel to the country and begin working with local people to get the technical job done, questions often arise regarding communication, cultural differences, leadership, and effective ways to work in groups. As one student participant has written, "these projects have presented me with excellent chances to grow as a leader. In recent trips, I have been entrusted with managing the on-site carpentry. In executing those responsibilities, I have learned that it is not enough to be [technically] skilled. A leader must be a communicator, a motivator, an advisor, and a teacher."[6]

Other questions generated by the international service experience involve the culture, language, and history of the Mayan people with which students are working. As one participating student has written, "I spent hours and hours pulling concrete as it came down the shoot from the concrete mixer, and that afternoon of solidarity stands out as my favorite time spent at the bridge site, working together as gringos, as Maya, as human brothers and sisters." Students begin to consider how the bridge will affect the community in the broadest sense. Since the bridge can provide a road over which building materials can be transported to construct the first community school, and can provide a shorter route to medical clinics and markets, the importance of appropriate infrastructure to education, health care, and economic development becomes evident.

Eventually, engineering students have more open-ended questions regarding the broad meaning of their work, their personal life and purpose, how to use their knowledge to meet the world's needs, and social justice. A recent engineering student commented that his bridge-building trip to Guatemala has affected him "both spiritually and emotionally, and in that regard, (related) classes . . . promise to fuel the fire of my passion for justice." And another writes that he and his companions "returned to the United States with broadened knowledge of bridge building and an inner commitment to help developing nations raise their quality of life." Another student similarly writes that "I believe I have a better perspective on Guatemalan culture and economics. More than this, I have a deeper understanding that the status of an area or people as underdeveloped is not as irrevocable as I first thought. I have been freed from the lie that I am powerless to affect a nation. This is one lesson I plan to take with me through the rest of my life."

Many students become more interested in international relations and the ethical relationship of rich and poor, and develop a desire to understand and embrace cultural differences. The experience becomes more enriching as these questions are discussed in classes pertaining to international development, engineering design, and appropriate technology. Students can also begin to explore their engineering work in terms of vocation, the good work they are called to do in life. For example, a student has written that "overall, this experience gave me the opportunity to reflect on how the use of my engineering skills could have an impact on the well being of others. After spending this time down in Guatemala, I plan on participating in future projects of this nature."

SERVICE LEARNING AND COMMUNITY-BASED RESEARCH

Service learning can be defined as "action and reflection integrated with the academic curriculum to enhance student learning and to meet community needs."[7] The approach has been formalized, and a mature pedagogy has been developed, although it is not yet widely applied to engineering education. One defining feature of service learning is that the experience involves both academic work for credit as well as service. For example, working in a community food pantry or tutoring in-need children is not service learning in itself, but working for a community and considering problems within university classroom discussion, reading, research, and design of appropriate technical solutions is service learning. This offers ethical situations and community interaction that enhances student reflection, helping students contemplate engineering solutions in a global and societal context. Service learners perceive an increase in their tolerance, personal development, and communication skills. Participants have reported being more curious and working harder.[8] As one student participant has written "I saw [Guatemalan] kids willing to work hard to achieve their goals, but opportunities for them to prosper from this hard work were minimal. Upon returning to the United States, I found myself eager to take advantage of any opportunity to better my future. I realized that the opportunities that emerge at Marquette are non-existent in Guatemala."

Service learning has been incorporated in some university engineering programs,[9] but the approach has been more widely applied to humanities and social science classes. Community-based research is a recent pedagogical extension of service learning. This approach

combines service learning with a collaborative process among professors, students, and community members in which research results are
used to enhance social justice.[10] Community-based research has not
yet been fully realized in engineering education, although at Marquette
University a community-based research engineering project in El Salvador is in progress.

PRESENT INITIATIVE AT MARQUETTE UNIVERSITY: HEILA

It may be most informative to perform community-based research
and service learning in a community far removed from the learner's
daily experience. In this way, knowledge of another culture may be
gained. Producing an engineering solution in the context of a different
language, culture, location, and economic framework can open the
mind to consider new technical and societal issues, while holding up a
mirror to the learner's own culture. In addition, graduates gain skills
and experience to become more effective when working in multi-
cultural teams and in foreign countries. Regarding civil and environmental engineering, Marquette University, Cornell University, Colorado
School of Mines, Michigan Technological University, and the University of Colorado, among others, offer international service learning.

As a Jesuit University, Marquette strives to expose students to the
promotion of justice and a preferential sharing with the poor. These
are directly related to engineering through development, sanitation,
clean drinking water, renewable energy systems, cleaner combustion
technologies, safe structures, and other engineering works that benefit
people in need. To join engineering, social science, and ethical discernment in engineering, Marquette University has been building the
Health, Environment, and Infrastructure in Latin America and Africa
(HEILA) undergraduate initiative since October 1999.

The initiative integrates an engineering capstone design course
and a seminar class with social science content. In the capstone class,
senior student groups consisting of three to five members work to
plan, analyze, and design solutions to community problems. Emphasis
is placed on engineering practice, technical communication, and pro-

fessional ethics. Student groups prepare both oral and written reports for a grade. The course also provides an opportunity to integrate previous course work to solve a practical engineering problem. Knowledge from existing engineering courses, such as "Water and Wastewater Treatment Plant Design," "Structural Analysis," "Bridge Design," and "Environmental Engineering," must be employed.

In the past, the capstone design was limited to large-scale, domestic projects identified by practicing engineers who served as mentors. Typically, a mentor guided students through a design with which he or she had already been involved from conception to design. There was no opportunity for students actually to construct the final product. Through the HEILA initiative, new, moderate-scale international projects have been identified so that students can construct bridges, drinking water systems, sewers, wastewater treatment systems, and other works for in-need communities in association with interested mentors and international sponsors.

The seminar class with social science content is titled Health, Environment, and Infrastructure in Latin America and Africa. The goal of the class is to help students contemplate the relationship between their major area of study and broader world issues, including history, culture, politics, social welfare, healthcare, and engineering infrastructure within developing and developed countries. Emphasis is placed on alleviation of poverty and developing nations. Subjects presented include (1) history and culture, (2) peace and justice issues, (3) appropriate technology for water treatment and sanitation, (4) the benefits of engineering infrastructure, (5) international healthcare issues, and (6) leadership. Students are asked to reflect on the importance of their major area of study in light of lecture and reading viewpoints, with the goal of gaining a richer understanding of the implications of engineering infrastructure in society. Class enrollment is not limited to engineering students. Participants from other disciplines, including physical therapy, international affairs, health sciences, and history often take the class.

Students are required to write a final paper for the seminar class, the major goal of which is to creatively tie together guest lectures, readings, service-learning experiences, and research to gain a deeper understanding of a specific topic and demonstrate an understanding of the class material. A draft of each paper is read by two students, and written reviews are provided to the authors. The student authors are then required to reply to each review, and, if necessary, incorporate changes in response to helpful criticism. Therefore, additional goals of the class are to become familiar both with peer review of another author's work and to learn how to use and respond to constructive criticism to improve a paper. Past student paper titles include "Privatization in Latin America: Economic Theory and Social Cost," "Drinking Water Quality and Health Concerns in the United States and Latin America," "The Ripple Effect of Service Learning," and "The Effect of Improved Transportation on International Development." The final papers are bound together in a volume of collected papers that is distributed to all class participants at the end of the semester.

It is important that students explore the relationships among class readings, class lectures, service-learning experiences, and their paper topics. Not all subjects lend themselves to synthesis of information from all sources. But students are encouraged to include as many sources of information as possible. Methods that students have chosen to present service-learning information include writing personal reflections, incorporating letters and discussions with people they worked with, and including digital images of important service-learning locations, events, and people.

SERVICE-LEARNING PROJECTS

During the last five years, Marquette engineering faculty members have mentored thirteen student senior design groups that designed or constructed a house and community center in El Salvador as well as five bridges, a sewer system, and a wastewater treatment plant in Guatemala. A biomedical engineering group has designed and built spirometers that were used in clinical studies of women garment workers in El Salvador, and mechanical and electrical engineering students have designed a solar energy system for a Haitian grade school.

It is important that students and faculty mentors teamed with groups rooted in the community to be served. To this end, HEILA engineering students work with nonprofit organizations, including Pro-

grama Ayuda para los Vecinos del Altiplano (PAVA) and Global Outreach Missions in Guatemala; Universidad Centroamericana (UCA)
and Movemento de Mujers Meleda Anaya Montes (MAM) in El Salvador; and the Coalition of Children in Need Association (COCINA) in
Haiti. The projects were identified by the local communities.

As an example of design/construction completed, a bridge project is described in the following paragraphs. The thirty-foot-long,
reinforced concrete bridge was constructed for the communities of
Comalapa and San Martin, Guatemala during 2002 winter break (January 10–22). The communities are within the south central highland region of the country in the Department of Chimaltanengo. Residents
requested the construction of a bridge over the Pichiquiej-Quemaya
River because the river flow is high during the rainy season that lasts
from October through March. During this season, crossing the river is
dangerous or impossible for the approximately twenty thousand in-
habitants. During the dry season, however, the river may be only a few
centimeters deep, and can be crossed easily (see figure 8.1). When the

Figure 8.1 Pichiquiej-Quemaya bridge site before construction during the dry
season, January 2002

river cannot be crossed, the trip between the two communities takes approximately three to four hours longer. Each community has a different public market day when people gather to buy and sell goods, and a bridge was desired, in part, to gain consistent access to both markets. Other advantages of the bridge include more rapid access to healthcare facilities and schools.

A group consisting of three Marquette University senior civil and environmental engineering students surveyed the bridge site in January 2002 (see figure 8.2). The survey was completed in one day. While surveying, students often have an opportunity to interact with local people as they learn about engineering. As one student commented "from the first minute we set up our surveying equipment, the entire neighborhood was perplexedly watching our every move, with the curious children asking questions and giggling as we measured our way. . . . Overcoming a number of unique challenges and setbacks, including encounters with mud, wild animals, and trouble keeping the rod functional, we remain hopeful that the collected data are sufficient to press forward this semester with a design that will ideally be imple-

Figure 8.2 Pichiquiej-Quemaya bridge site student survey, January 2002.

mented over the next year. There's no better way to learn than through the unexpected challenges faced in a developing country."

The remaining twelve days of the student's Guatemala visit were spent building a previously designed bridge in San Martin with a group of approximately fifteen other North Americans and thirty Guatemalans. This other bridge had been designed the previous year by other students. The work gave students the opportunity to learn appropriate construction methods, including site excavation, form work construction, concrete mixing and placement, as well as reinforcing steel preparation. These activities for a similar project have been described in detail elsewhere.[11] Upon their return to the United States, the students designed a bridge and prepared a written report in the capstone class. An oral presentation was made for the class, including both technical details and cultural experiences. The students then graduated in May 2002.

The participants were also able to learn about the Guatemalan highland culture. For example, when trying to communicate with local children, students discovered that, although the official language is Spanish, most of the local people spoke Kaqchiquel. The students performed research and reading after their return, and discovered that approximately twenty other indigenous languages are spoken in Guatemala. About 55 percent of the Guatemalan population is Native American and of Mayan decent, whereas 44 percent is of mixed Native American and Spanish origin, also called mestizo. In contrast, the Native American population in other American countries is much lower. For example, in El Salvador the population is only 1 percent Native American.

In January 2003, some members of the design group returned to Guatemala as alumni and helped construct the bridge they designed the previous year (see figure 8.3). Therefore, they were able to experience the design and construction process from surveying to finished product. This pattern is typical in that most students want to return to the foreign country after graduation to help build what they have put on paper. As a student has written, "I hope to have the opportunity to return to Guatemala next January to participate in the construction of the bridge I designed."

Figure 8.3 Concrete deck formwork for Pichiquiej-Quemaya bridge, January 2003.

THE VALUE OF SERVICE LEARNING

International service learning provides a unique opportunity within the undergraduate engineering education curriculum in that it not only facilitates a design and build project but also provides a mechanism by which students are exposed to social science issues and fosters an understanding of ethical discernment. Too often engineering students develop an attitude of providing technical solutions while overlooking the broader social, ethical, economic, and quality of life issues associated with their design. An international service learning experience exposes students to a cross-cultural environment and forces them to consider the technical challenges associated with communication, resource allocation, appropriate design, and possibly social justice issues. This process often causes students to reflect on the impact of an engineering solution on an environment and seek a broader understanding of their role within society. Designing viable engineering solutions within this framework can not only challenge and grow a student's technical expertise but also stimulate a sense of empowerment, raise their level of social awareness, and instill the significance of ethical discernment.

The Marquette service-learning project has been crafted to incorporate both a directed classroom experience and a hands-on international design and build experience. The goal of the classroom experience is to help students contemplate the relationship between their major area of study and broader world issues, including history, culture, politics, social welfare, health care, and engineering infrastructure within developing and developed countries. The goal of the hands-on experience is to provide students with a tangible working example of an engineering project from design to construction. Together through these two components, the educational experience can take on a new and richer dimension as students begin to understand engineering as a vocation. These experiences are especially appropriate when teaching technical skills while integrating Catholic beliefs and ideas, such as the service of faith and promotion of justice rooted among the poor, into engineering education.

NOTES

The authors thank the Marquette University College of Engineering, including the Departments of Civil and Environmental, Mechanical and Industrial, and Biomedical Engineering, and the Marquette University Water Quality Center for providing travel funds to help support the projects described.

1. P. Vesilind, "Engineering as Applied Social Science," *Journal of Professional Issues in Engineering Education and Practice* 127, no. 4 (2001): 184–88.

2. Engineering Accreditation Commission, *Criteria for Accrediting Engineering Programs* (Baltimore, MD: Engineering Accreditation Commission, 2000).

3. W. Wolf, "The Social Responsibility of Engineers (and Its Implications for Engineering Education)," The George W. Woodruff School of Mechanical Engineering Annual Distinguished Lecture (Georgia Institute of Technology, Atlanta, 2000).

4. M. Valenti, "Teaching Tomorrow's Engineers," *Mechanical Engineering* 118, no. 7 (1996): 64–69.

5. C. Lowney, *Heroic Leadership* (Chicago: Loyola Press, 2003).

6. This and subsequent quotes from students about their experiences are from anonymous course evaluations. Names and dates were confidential and hence cannot be reproduced here.

7. P. Wankat and F. Oreovicz, "Learning Outside the Classroom," *ASEE Prism* 10, no. 5 (2001).

8. J. Eyler, D. Giles, and J. Braxton, "The Impact of Service-Learning on College Students," *Michigan Journal of Community Service Learning* 4, no. 4 (1997): 5–15.

9. D. Zitomer, M. Gabor, and P. Johnson, "Bridge Construction in Guatemala: Linking Social Issues and Engineering," *Journal of Professional Issues in Engineering Education and Practice* 129, no. 3 (2003): 143–50; D. Zitomer and P. Johnson, "International Service Learning in Environmental Engineering," in *Proceedings of American Society of Civil Engineers World Water and Environmental Resources Congress* (Reston, VA: American Society of Civil Engineers, 2003); D. Vader, C. Erikson, and J. Eby, "Cross-Cultural Service Learning for Responsible Engineering Graduates," in *Proceedings of American Society of Civil Engineers Education Conference* (Reston, VA: American Society of Civil Engineers, 1999); E. Coyle, L. Jamieson, and L. Sommers, "EPICS: A Model for Integrating Service Learning into the Engineering Curriculum," *Michigan Journal of Community Service Learning* 4, no. 4 (1997): 81–89; E. Tsang, C. Martin, and R. Decker, "Service Learning as a Strategy for Engineering Education in the Twenty-first Century," in *ASEE Annual Conference Proceedings* (Washington DC: American Society for Engineering Education, 1997).

10. R. Stoecker, "Community-Based Research: From Practice to Theory and Back Again," *Michigan Journal of Community Service Learning* 10, no. 4 (2003): 35–46.

11. Zitomer, Gabor, and Johnson, "Bridge Construction in Guatemala."

PART 4

Formation and Preparation of Students

Chapter Nine

Vocational Awareness in Engineering Students

SCOTT J. SCHNEIDER

As a faculty member I am excited about not only teaching the technical aspects of the engineering curriculum, but also about helping students realize their vocational callings to the engineering profession. While working in industry as an engineer, it took me many years to start understanding engineering as my vocation, how God has called me to serve him. This realization has changed my perspective on the engineering profession and how it relates to helping spread God's word through the actions of its members. As a professor, I hope to foster this vocational calling in my students early on so they can enjoy the great rewards that come from serving God through the use of their skills.

It is my honest belief that one's skills are gifts from God, not to be used for selfish reasons but to be shared through helping others. The skills associated with the engineering profession are so critically needed in today's world. Often, however, the largest engineering research and development efforts tend to be focused on destructive rather than constructive purposes. Perhaps because of this, the engineering profession is often viewed as a rather sterile manipulation of facts, figures, and equations towards solutions to technical problems. It would appear that little attention has been placed on the engineering profession as a method of provoking social change or providing solutions to humanitarian crises. My mission, therefore, through this research effort is to define methods that can change this perception by focusing on developing the engineering students' social and faith awareness.

God has helped place each student in the University of Dayton's School of Engineering program. Some students within the program are intent on discovering the true meaning of their vocation as future engineers; others are not as in-tune with what vocation is, or even where God fits in their life. There are currently several focused opportunities for engineering students to explore their engineering vocation in relation to their own social and faith-based values. These opportunities, however, mainly reach those students already aware of their vocational calling and who are actively trying to create synergies between their faith and their skills as future engineers.

I am interested in defining how engineering educators can reach a larger base of students in a meaningful exploration of their vocation as it relates to their individual faith and values. I want to focus my research on developing successful methods for broadening vocational awareness in all engineering students. The first step in this research effort is to investigate in what activities professors are currently engaged to help promote such awareness. This paper provides an initial overview of results from a survey of engineering professors at both secular and faith-based universities across the United States.

ENGINEERING AS VOCATION

The word vocation has several meanings as found in any dictionary. The *American Heritage College Dictionary* defines vocation as "1. A regular occupation, especially one for which a person is particularly suited or qualified. 2. An inclination, as if in response to a summons, to undertake a certain kind of work, especially a religious career; a calling."[1] Though several forms of technical education have been classified as vocational, the definition used for this research and in this paper is most closely tied to the second definition cited above.

So, what is a vocational engineer? My personal definition of a vocational engineer are persons who are passionate about their profession. They not only recognize the need to perform their job within an ethical and moral framework, they also realize that they are obliged to do so. Vocational engineers are ready to use their skills however God calls them to do so. Most importantly they recognize that their engineering skills are gifts and need to be shared to help others, not used only for selfish reasons.

Vocational Foundation of Engineering

The foundation of the engineering profession is important in that it helps signify the true intentions of the discipline, as well as its impact on society and on the human condition. Engineering as a vocation is not necessarily a new idea but perhaps is a relatively new phrase. Engineering is seen by many scholars as being born out of the military and its need for technological advances for security and conquest. Such a foundation, however, does not align the beginnings of the engineering profession with the Christian moral principles necessary for it to be considered a vocation.

Upon further review of the historical foundation of the engineering profession, Brad Kallenberg investigates in his chapter "The Theological Origins of Engineering" other historical roots of the engineering discipline that provide a better footing for the profession as a Christian vocation. He traces engineering back to the monastery of St. Victor in Paris where Hugh provides perhaps the first theological support of a forebear to the engineering profession, the mechanical arts.[2] More importantly, Kallenberg draws upon Hugh's theological accounts to demonstrate that the technological advances made possible through the engineering profession can actually aid humans in their journey towards reunification with God.

Today the notion that the engineering profession can be a vocation is starting to crop up at numerous Christian universities and colleges across the United States. In her essay "Reflections on Integrating Engineering into My Christian Life," Gayle Ermer, an engineering professor at Calvin College, mentions that engineering, through the creation of technology, can be used as a tool to help satisfy the needs of humans as mentioned in Matthew 25: "For I was hungry and you gave me something to eat, I was thirsty and you gave me something to drink, I was a stranger and you invited me in, I needed clothes and you clothed me, I was in prison and you visited me, I tell you the truth, whatever you did for one of the least of these brothers of mine, you did for me."[3] She goes further to imply that through faithful service to God an engineer will also be upholding the expected ethical responsibilities of their profession. Ermer also supports the notion that technology is not value-free; instead, it can have a significant impact on humans and society. She therefore calls for the development of culturally appropriate technology.

In support of Ermer's position on engineering as a Christian vocation, Daniel Lynch states in his paper "Catholic Social Thought and Engineering Education" that "it seems Jesus had an engineering background (carpentry) which ought to be mirrored in today's profession: feed the hungry, clothe the naked, etc. To the extent that our profession focuses on these elementary imperatives, we have a good profession."[4] It is important to note the commonality in views of engineering as vocation from professors at different universities. Lynch draws upon this central theme further by imposing upon humans the current material conditions that exist and the associated need for engineers to work on improving this condition and how by doing so they will serve and glorify God.

ASSESSING ENGINEERING FACULTY SENSE OF VOCATION

Given a defined notion of vocation for engineers, a question must be asked. Do engineering faculty at a Catholic University see their role as vocational? To answer this question, a survey was created to assess the embrace of a vocational perspective of the faculty at Catholic and secular universities. In constructing this survey on methods for promoting engineering as a vocation, I found two approaches that I could take. One approach would be to formalize the survey so that the results could be statistically analyzed. The second approach would be more concerned with obtaining personal stories, advice, and insights into the individual respondents' understanding of engineering as a vocation and how they introduce this understanding to their students. I decided to take the second approach, because the nature of this research is not to calculate statistical results but rather to help develop meaningful methods for engaging students' awareness of engineering as vocation.

Sixty engineering faculty members responded to this survey. From those who responded, twenty-five teach at secular universities and the rest teach at faith-based universities. Most of the responses are meaningful and provide insight into the common understanding of engineering as vocation and how one might go about teaching that to students. The following subsections look at each portion of the survey individually, drawing conclusions, and incorporating faculty comments.

Personal Beliefs about Engineering as a Vocation

The first section of the faculty survey is intended to discern what the phrase "engineering vocation" means to the individual professors. I chose not to include my perspective of engineering vocation so as to not sway the respondents' opinion. I am mostly concerned with their initial, unbiased views. Only two respondents used the secular interpretation of vocation as simply a professional occupation.

Aside from trying to discover the methods professors are currently using to "teach" engineering vocation, there is also interest in discerning the differences between how this subject is understood between professors at secular and faith-based universities. The two questions posed in this section of the faculty survey are:

1. Do you see engineering as a vocation, and if so what does this mean to you?

2. Is having a religious faith required for a vocational understanding of an occupation such as engineering?

One statistic gleaned from the responses to these questions is that 68 percent of the respondents that teach at a secular university are aware of the concept of engineering vocation; however, due to the lack of any explicit expression of faith on their campus, they are only able to approach it as engineering professionalism or engineering ethics. Therefore, I see a major advantage able to be played out in faith-based institutions compared to secular institutions. For example, Lynch points out that "many approaches to professional ethics skip the essential starting point and easily convey to students a world of rule-making which is strictly utilitarian, without principled direction or compass. Catholic social thought injects the relationship between God and man as a point of order and direction."[5]

Engineers are always eager to understand the "why" behind what they see and learn. When dealing with the differences between engineering professionalism and engineering vocation there is a major distinction on how the "why" is addressed, or even addressable at all. The engineering professionalism camp must rely on concepts from secular humanism that rely on the pervasiveness of human morality to direct our actions and decision-making capabilities. In contrast, engineering vocation as taught at a Catholic university can rely, for example, in defining the "why" on the Catholic social justice doctrine.

From the 72 percent of respondents who did see engineering as a vocation, two major themes appeared in their responses: individuals with certain talents or gifts are predisposed to the engineering profession and to utilizing their skills for the service of others. While professors coming from a more secular viewpoint may not have drawn the conclusion of how these notions can be supported using a God-centered viewpoint of vocation, they could draw, I believe, a similar conclusion fairly easily. However, only fifteen respondents felt that a religious faith is critical in developing a vocational awareness of engineering. The majority of respondents from both the secular and faith-based universities instead pointed to an internal ethical drive that could lead an engineer towards a vocational awareness of their profession, a point that is supported by secular humanism. There are two points that can be drawn from such a conclusion. First, by not requiring a religious faith for understanding and even practicing engineering as a vocation, these respondents are not alienating students who do not practice a religious faith. Secondly, by not requiring a religious faith for vocational awareness, professors at faith-based universities are perhaps missing an opportunity to draw synergies between their institutions' faith perspective and the discipline being taught.

Professional Beliefs about Engineering as a Vocation

The primary goal of all students passing through university engineering programs is to gain the necessary skills to enable them to obtain a satisfying position within industry or to continue their studies at the graduate level. The focus of the next section of questions of the faculty survey is to investigate the professional consequences of a student's vocational awareness as a future engineer. The questions posed during this section of the faculty survey are:

1. How does an engineer who sees his occupation as a vocation perform his job differently than one who does not?
2. What impact can a vocational engineer have in the workforce, and is a vocational engineer marketable?

The majority of respondents found that vocational engineers would have a higher degree of professionalism and would be more inclined to approach their career from an ethical perspective. One respondent from a secular university wrote "anyone who feels that they are called, by whatever deity they believe in, to perform a particular career is likely going to perform that career with higher dedication, deeper conviction, greater concern for the impact of that career, and more consideration of those who are effected by it. Engineering would be no different than any other career in this regard." Surprisingly, however, only three professors commented on the likelihood of a vocational engineer explicitly to seek out engineering activities directly related to the service of others, a concept that seems intrinsically tied to the notion of a vocational engineer.

Another interesting point made by one respondent is that "some-one who sees their occupation as a vocation feels a responsibility to feed the young (i.e., mentor the new members in the occupation) and they feel a responsibility to contribute to the profession (i.e., publish, present, etc.)." This response is unique and, in a different light, does embody the Christian notion of vocation as used in this research. Vocational engineers are not only obliged to share their gifts for the better-ment of others in society as a whole, but also for the betterment of the professional community. The implications of such a position on the en-gineering profession are huge. Engineering students who leave the universities with a keen awareness of what it means to be a vocational engineer, especially those with a religious faith perspective, are also the engineers who help build the body of knowledge that is the foundation of the profession. When professors emphasize the vocational aspects of engineering to their students, they are in turn also helping mold the future directions of the profession and its role in society. Other re-spondents supported this position by repeating the notion that a voca-tional engineer reinforces the good nature of the profession and like-wise promotes a better public perception of engineering.

Overall, the responses converged towards defining a vocational en-gineer in industry using such words and phrases as motivated, ethical, God-centered, not focused on pay but on project, role-model, profes-sional, quality of workmanship, humbleness, and purposefully driven. A few respondents felt a vocational perspective of engineering would be a hindrance in the current profit driven marketplace; the traits listed above, however, would seem to indicate otherwise. The responses in this section of the faculty survey definitely support the importance of motivating all engineering students towards a vocational perspective of their future profession.

How You Teach Engineering as a Vocation

The most important section of the faculty survey focused on the actual implementation of engineering vocation into the curriculum. Within this section, the questions raised are focused on revealing what faculty members have used to motivate vocational awareness within their students and the impact of this effort. I am also interested in discerning what impact religious faith has in the ability to promote engineering vocation. The questions asked during this section of the faculty survey are:

1. Do you openly try to foster the concept of engineering as vocation for your students, and if so how?
2. What is the best mechanism to help students understand the vocational aspect of engineering?
3. Can engineering as vocation be taught, and if so, can it be taught from a secular perspective?
4. Do you see students' understanding of engineering vocation as being an important aspect of their education, and if so why?

Of the sixty respondents, thrity-five indicated that they in some way try to foster vocational awareness in their students. The awareness that they are fostering, however, depends upon their own understanding of what it means to be a vocational engineer. Twelve respondents try to foster the perspective that engineering as a vocation is rooted in a deeper calling. These respondents utilized two major styles of fostering this vocational awareness: implicit methods and explicit methods. The faculty members using implicit methods rely on the example they set through their interactions with the students and with respect to their professional commitments. The second group, those who explicitly try to foster vocational awareness, tend to utilize two main methods for encouraging students' awareness of engineering as a vocation: engineering volunteerism or an engineering ethics course. No respondent mentioned a method used throughout a specific department, their school of engineering, or their university related to aiding a student in their understanding of engineering as a vocation.

Engineering service or volunteerism is a growing trend that reinforces the critical, often key, community service nature of the vocational engineer. Service learning within engineering programs got its start through a philosophy of experience in education from the work of John Dewey.[6] According to William Oakes, there are four key ingredients to service learning: service, academics, partnerships, and analysis and reflection.[7] Today's service-learning exercises are often rooted not only in the service provided, but also in the concept of appropriate and sustainable technology. Appropriate and sustainable technology, as defined by Carl Eger and Margie Pinnell, involves solutions to problems using alternative, nontraditional technologies that are based on fundamental science and engineering principles, are culturally appropriate, can be made and maintained by the local people, and promote self reliance and help feed the local economy.[8] Through service-learning activities, especially those reinforcing appropriate and sustainable technology, students not only hone and develop their engineering skills, they also serve the local and global communities in which they work. Service learning is a key ingredient in the formation of any vocation.

At both secular and faith-based universities, an engineering ethics course has been a requirement in most engineering programs for a long time. These courses typically rely on case studies to help demonstrate the importance of using ethical and professional behavior throughout the decision-making and design processes associated with the engineering disciplines. Religious faith can be used successfully within an ethics course to help supplement the more traditional philosophies covered. Robert Barger has developed a computer ethics course at the University of Notre Dame in which he utilizes Catholic social thought to augment the four major ethical philosophies: idealism, naturalism, pragmatism, and existentialism.[9] His method does not replace these philosophies but demonstrates how the social teachings of the Catholic Church can be used to highlight errors within these philosophies. Barger also demonstrates how Catholic social thought can be used within engineering to deepen students' understanding of commitment and ethical responsibility.

There are other methods mentioned by single respondents with respect to developing student awareness of engineering as a vocation. These methods include prayer in class, periodic thoughtful reflection sessions within class, and the use of a student vision statement. The impact of these methods on influencing a students' vocational awareness may be less certain than the impact caused through engineering volunteerism projects or through an engineering ethics course. These methods, however, may provide necessary "glue" within a curriculum to tie together the engineering service projects and ethics course outcomes. The idea of having freshman students develop a vision statement that includes not only professional but also spiritual and personal milestones could be a logical foundation in any curriculum wishing to develop a more cohesive approach to teaching vocational awareness.

Many respondents found that the action of encouraging students to explore their vocational awareness of engineering is more tied to inspiration than teaching. There is an almost equal split in respondents who felt that vocation is able to be taught (inspired) from a secular perspective and those who felt that it must be promoted in a religious faith context. Again, a division appears between those who see a vocational engineer as synonymous with a professional engineer and those who see it as a deeper calling from God.

INCORPORATING RESULTS INTO THE CURRICULUM

While no respondent in the faculty survey delivered a complete package that can be implemented throughout a department, school, or university program to foster students' vocational awareness, many respondents have used several mechanisms very successfully, mechanisms that could be pulled together to develop a more formal and cohesive approach. Such an integrated approach could, for example, utilize service learning (engineering volunteerism) within capstone projects and locate the contents of an engineering ethics course deeper into the curriculum with follow-up case studies and paper reflections in other courses. Several universities have already started stepping away from teaching ethics as a single course to a more holistic approach where the students continually receive ethics-related material throughout their studies. It is believed that the greater the variety of ethics courses an engineering student is exposed to, the greater the impact, especially when the individual ethics components are joined with their logical technological partner (that is, data collection and interpretation with integrity taught in a laboratory course).[10] In addition to these two very powerful tools, the power of prayer and faith-based reflections within a course should not be overlooked. The true testament to how effective any of these methods are is to ask the students themselves. A student survey is being administered to a selected group of students at several universities who have expressed a deep appreciation for the vocation of engineering. The results from the student survey will help clarify what methods have the greatest impact and then provide some final form to a unified approach that can be established throughout a curriculum to help motivate students to explore their own awareness of engineering as a vocation.

Facilitating Change

There are many positive examples at the University of Dayton and other universities across the nation indicating the desire of many professors to help strengthen their engineering programs' emphasis on teaching the whole person through vocational awareness. Such examples at Dayton include the ETHOS service-learning organization and an associated course on sustainable and appropriate technology, support for research such as this essay, the development of this conference on the Role of Engineering at a Catholic University, and a strong emphasis on engineering as vocation from the new dean of the School of Engineering. Even with all of these advancements much still needs to be accomplished.

A follow-up faculty survey has been sent out to those faculty members who are actively promoting engineering vocation within their curriculum. The focus of this follow-up survey is to help understand some of the political issues within their individual schools related to this topic. One of the key issues is how the School of Engineering as a whole supports research related to engineering as a vocation, service learning, or time spent in the development of engineering ethics material. This support is not just financial funding from the School of Engineering, but also how these efforts are recognized within the engineering faculty community and supported throughout the promotion and tenure process. Furthermore, it is important to look at how the vocational awareness of the faculty and the engineering program as a whole is advertised to incoming students. Those prospective students visiting an engineering program committed to promoting the vocation of engineering should have a clear realization that this School of Engineering is different, and they will learn not only with their minds, but also their hearts and souls.

Perceived Obstacles

A commonly perceived obstacle to teaching engineering as a vocation came up in the faculty survey over the demographic of the classroom: namely, the fact that students may have varied faith backgrounds or no faith at all. Moreover, some students are perhaps not destined to work as engineers professionally and will not therefore see it as their calling. In response to such issues I look to the parable of the sower told by Jesus in Matthew 13:3-9: "Then he told them many things in parables, saying 'A farmer went out to sow his seed. As he was scattering the seed, some fell along the path, and the birds came and ate it up. Some fell on rocky places, where it did not have much soil. It sprang up quickly, because the soil was shallow. But when the sun came up, the plants were scorched, and they withered because they had no root. Other seed fell among thorns, which grew up and choked the plants. Still other seed fell on good soil, where it produced a crop—a hundred, sixty, or thirty times what was sown. He who has ears, let him hear.'" The seed is the Word of God spoken to us. Likewise, as educators if we work to reveal a vocational understanding of engineering in our students, we may not reach all of our students, but those that are reached will in their work as a vocational engineer help spread God's word and our efforts will be multiplied.

NOTES

This research has been supported by the University of Dayton's Faculty Fund for Vocational Exploration through a grant from the Lilly Endowment, Inc.

    1. *American Heritage College Dictionary,* 3rd ed. (Boston: Houghton Mifflin Company, 1993), 1511

    2. Brad J. Kallenberg, "The Theological Origins of Engineering," chap. 2 of this volume.

3. Gayle E. Ermer, "Reflections on Integrating Engineering into My Christian Life," Calvin College Faculty Statement on Integration of Faith and Learning, September 2002.

4. Daniel R. Lynch, "Catholic Social Thought and Engineering Education," (paper presented at Catholic Social Thought Across the Curriculum Conference, University of St. Thomas, Saint Paul, MN, October 2003).

5. Lynch, "Catholic Social Thought and Engineering Education."

6. William Oakes, *Service–Learning in Engineering: A Resource Guidebook* (Campus Compact, 2004), 6.

7. Oakes, *Service-Learning in Engineering, A Resource Guidebook*, 7–8.

8. Carl W. Eger and Margaret F. Pinnell, "Appropriate Technology and Technical Service in Developing Countries (ETHOS) Elective Course" (paper presented at the 2005 American Society for Engineering Education Annual Conference, Portland, OR, June 2005).

9. Robert N. Barger, "Catholic Social Thought and Computer Ethics" (paper presented at Catholic Social Thought Across the Curriculum Conference, University of St. Thomas, Saint Paul MN, October 2003).

10. David A. Rogers, "Work in Progress–Ethics Integrated into Engineering Courses" (paper presented at the 34th ASEE/IEEE Frontiers in Education Conference, Savannah, GA, October 2004).

Chapter Ten

Helping Students Discern Engineering as a Vocation

CARMINE POLITO, DOUGLAS TOUGAW,
AND KRAIG J. OLEJNICZAK

OVERVIEW OF VALPARAISO UNIVERSITY

Valparaiso University is dedicated to superior teaching based on excellent scholarship. As a scholarly community it actively engages in the exploration, transmission, and enlargement not only of knowledge but also of the cultural and religious heritage of human society. It is proud to prepare men and women for professional service. This community values respect for learning and truth, for human dignity, for freedom from ignorance and prejudice, and for a critically inquiring spirit. The university aims to develop in its members these values, together with a sense of vocation and social responsibility. It holds that these values receive their deepest meaning and strength within the context of the Christian faith. These basic commitments enable Valparaiso University to graduate students whose individual achievements and aspirations are linked invariably to larger social, moral, and spiritual horizons of meaning and significance. Proud of all its alumni who have carried its values into leadership roles in their communities, the church, social institutions, the nation, and the world, it aims to continue graduating such potential leaders.

Since its inception over 150 years ago, the university has passed through three distinct phases. Begun by Methodists in 1859 as an institution pioneering in coeducation, the Valparaiso Male and Female College was forced by the reverses of the Civil War to close its doors in 1871. An enterprising educator, Henry Baker Brown, revived it in 1873 as the Northern Indiana Normal School. "Mr. Brown's School," a flourishing private, proprietary institution, was renamed Valparaiso College in 1900 and rechartered as Valparaiso University in 1907. During the next twenty years, it won national recognition as a low-cost, no-frills institution of higher learning that served thousands of students who might not otherwise have been able to afford a good education. Many alumni from this period achieved distinction in their fields as governors, legislators, scientists, business leaders, and other professionals. However, after World War I the university went into decline and bankruptcy; then, in 1925, the Lutheran University Association purchased it, beginning the modern phase of the University's history.

The university's concern for the personal and intellectual development of each student is rooted in its Lutheran heritage. This Christian philosophy of education guides both the design of its curriculum and the approach to learning that it fosters. Beyond the courses in theology that the curriculum provides, the University emphasizes a Christian freedom that liberates the scholar to explore any idea and theory, a vocation freely uniting faith and intellectual honesty. In its residential life the university leads students to accept personal responsibility for their development and encourages a sense of caring for one another. Standing together at the center of the campus, the Chapel of the Resurrection and the Christopher Center for Library and Information Resources express the university's belief in the creative relationship between faith and learning. The university's motto, too, emphasizes this relationship: *In luce tua videmus lucem,* "In Thy light we see light."[1]

This motto, which succinctly encompasses the relationship between faith and learning, can also be linked metaphorically to Truth, and the role of a Christian university in the formation of men and women grounded in the Truth. In his inaugural address, entitled "The

Destiny of a Christian University in the Modern World," on the afternoon of October 6, 1940, the Reverend O. P. Kretzmann eloquently spoke about the theme of Truth:

> Essentially a University is a voluntary association of free men and women in a community which is dedicated to a two-fold task: the search for Truth and the transmission of Truth, free and unbroken, to those who are born later in time. Its first and supreme requirement is a company of men and women who will know Truth when they meet it, no matter whence it comes or whither it leads; who will love Truth more than riches and power; who will conduct the search for Truth with radical sincerity, intellectual honesty, and a deep reverence for even its smallest and faintest gleam; and who will be able to transmit this devotion to the generation who will live long after they have joined the company of the wise and the silent in the graveyards of the world. Especially in the modern world it must be the destiny of a Christian University to cling to the reality of universal truth. There is a moral, philosophical, and scientific Truth which must be one and the same for all races and all nations. For the modern heresy of the relativity of all standards it must substitute the concept of an order of absolute Truth, of absolute ethical goodness, of absolute social justice to which all differences must be submitted, and by which they must be judged. Although we must be ready at all times to admit the partiality of our apprehension of Truth, we must also stand sharply and immovably against the unintelligent and unreasonable pretensions of the philosophies in the modern world which identify the extent of Truth with our partial apprehension of it, or confine it to a certain race or nation. In the fullest and highest sense of the words the Christian University can be and must be, the most catholic and universal and democratic institution in the modern world. It can never compromise with the moral disorder of liberalism or the dangerous heresy that Truth is a slave not a master.[2]

Valparaiso University continues to embody this theme through its mission statement: "Valparaiso University, a community of learning dedicated to excellence and grounded in the Lutheran tradition of scholarship, freedom, and faith, prepares students to lead and serve in both church and society."

The hallmark of Valparaiso University's College of Engineering graduates has been their ability to *lead* and to *serve* in professional, community, and congregational settings. Since the College's founding in 1917, it has instilled these values through a rigorous curriculum that seeks resonance between the professional programs and the liberal arts. This integrative approach has produced graduates who have applied scientific knowledge, ethically and creatively, to benefit society.

The College of Engineering builds on three themes that form the unique Valpo experience.

1. *Engineering as a vocation.* The Christian character of our institution, when combined with the strong undergraduate education within the College of Engineering, creates a positive environment that is unusual in undergraduate engineering colleges. The college seeks to instill into its graduates a firm dedication to engineering ethics, an awareness of social justice, and an understanding of their responsibilities for environmental stewardship, resulting in a sense of purpose and personal mission.

2. *Engineering in a collaborative environment.* For nearly ninety years, one defining characteristic of the college has been the camaraderie among faculty, staff, and students. This supportive environment promotes a collaborative rather than a competitive learning atmosphere. Valparaiso University has a particularly strong cohesion of academic units across campus due to the purposeful integration of the professional and liberal arts programs. We are thus perfectly suited to continue to be a pioneer in what will certainly become the norm in higher education: collaborative, interdisciplinary, student-centered academic programming. Valpo will provide a home for fields of emerging technology that require diversity of expertise, drawing on the confluence of engineering, the life sciences, business, and the humanities.

3. *Developing the leaders of tomorrow.* Valparaiso University has always prepared strong leaders to fill the needs of society. Society will have an increasing critical need for values-based leaders who are technically skilled and have an understanding of the business world. Our

students will develop a clear personal vision, interpersonal skills, and the desire necessary to be effective leaders.

These three pillars support the engineering experience at Valpo. Along with a strong foundation of fundamental skills in mathematics, science, and problem solving, they are characteristics that enable our alumni effectively to lead and serve their society through their chosen vocation.

INSTITUTIONAL INITIATIVES

The concept of vocation is an important one in the Christian tradition, especially within the Catholic and Lutheran churches. Only recently, secular institutions have begun to recognize the importance of discernment, vocation, and pursuing a divine calling to a career other than that of a priest or minister. Of particular interest, several authors have recently begun to introduce the idea of a vocation or calling to professional careers in engineering and business.[3] Such a renewed focus on vocation is, at least in part, a reflection of the desire each of us has to know that the work we are doing is part of a larger plan and is contributing to a larger good. As a faith-based institution, Valparaiso University is dedicated to helping its students in all disciplines to develop precisely such an awareness of their place in God's larger plan.

Valparaiso University sponsors a wide variety of programs designed to help students deepen their sense of vocation to a particular field of work. This effort is coordinated by the Project on Theological Exploration of Vocation, which was funded by a $1.9 million grant from the Lilly Endowment beginning in 2001. The activities organized as part of this project are designed to "encourage students to consider how their lives and work may be informed by their deepest beliefs and conducted in service to others." It also focuses on faculty and staff, encouraging them to "develop theological perspectives on vocation in their teaching and in their own lives."[4] Although some aspects of the project focus on church vocations—the preparation of students who plan a career of full-time church service—the concept of vocation is applicable to every field of study. Students who are endowed with the appropriate skills may have a strong and worthy calling to a life of service in engineering as well as business, teaching, law, or medicine. As a comprehensive Lutheran institution, Valparaiso University strives to help every one of these students to discern his or her true calling and live a more meaningful and God-centered professional life.

The university's focus on discernment and vocation begins not with the students' discernment efforts but with those of the faculty. The institution has established a number of mechanisms to help both new and experienced faculty to discern their own calling to a life of teaching and scholarship and to understand how such a calling can affect their interaction with their students and colleagues. Faculty are also given the tools necessary to help students to discern their own vocation as part of the close interaction between faculty and students in the classroom and in advising sessions.

As part of the Project on Theological Exploration of Vocation, the university sponsors an annual ten-day seminar intended primarily for new tenure-track faculty members. This seminar "introduces theological perspectives on the vocation of a teacher-scholar at a university associated with the Lutheran Church."[5] It is offered to all tenure-track faculty at the conclusion of their first year of teaching at the university. The participants in this seminar travel to the Valparaiso University Study Centre in Cambridge, England, where they participate in twice-daily sessions exploring the ideas of Christian vocation, the vocation of teaching and scholarship, and the vocation of working at a church-related university.

Both new and veteran faculty members are encouraged to apply for funds to support the development of new courses and the strengthening of existing courses to incorporate explicitly the theological exploration of vocation into curricula across campus. Faculty from the liberal arts, sciences, and the professional colleges are all able to investigate the idea of vocation within the context of their specific discipline, helping students to see that discernment is not only a process intended for those considering a career as a clergy member, but also for those planning to become engineers, scientists, and nurses.

One of the most important opportunities a member of the faculty has to interact with students is by acting as an academic advisor. It is understood that academic advisors will thoroughly understand the details of the curriculum, helping advisors to provide students with sound advice about course selection in each semester, but advising can and should be about much more than selecting courses. It is also an opportunity to talk to younger students about the nature of their calling to the profession, and an opportunity to help more advanced students make career decisions that will support their own personal calling. In order to provide faculty with the tools they need to be such discernment-oriented advisors, the university sponsors workshops on vocation for all academic advisors. Focused on carefully selected readings and presentations by experienced advisors, the workshops focus on the Christian theological conceptions of vocation. The goal of the workshops is to prepare advisors to answer the wide range of questions students are likely to ask as they begin to think "in a meaningful way about their future work and life commitments."[6]

CURRICULAR INITIATIVES

Perhaps the most effective opportunity faculty have to affect the attitudes of their students about vocation is by integrating these ideas directly into courses and curricula. There are several opportunities for engineering faculty to help students gain a deeper appreciation for the importance of discernment to the successful selection of a rewarding career path. These opportunities begin in the first semester a student arrives on campus, and they extend all the way to his or her last semester as a graduating senior.

The first opportunity to discuss vocation occurs in the Valpo Core, which is a two-semester interdisciplinary course sequence required of all students in their freshman year. Core class sizes are intentionally kept small in order to promote more personal interaction among the students and between the students and the professor, creating an intellectually stimulating environment. The course sequence is composed of six units, each of which focuses on one aspect of human existence. The fourth unit, entitled "Work and Vocation," focuses on the theological perspective of vocation, providing students with the basic vocabulary necessary to understand campus-wide discussions of vocation and discernment. In order to ensure that the faculty are sufficiently knowledgeable to effectively lead such discussions, they each participate in an annual two-day seminar that helps them to develop teaching strategies and approaches to the texts selected for that unit. In this seminar, faculty examine the relationship between the Christian understanding of vocation and non-Christian concepts of duty and divine calling, such as that found in the *Bhagavad-Gita*.[7] The faculty also consider the ideas associated with vocation introduced by such authors as Frederick Douglass, Willa Cather, and the philosopher Plato.

At the same time as first-year students are being exposed to the basic ideas of vocation in the core sequence, they are also being asked to begin to consider their own professional calling in the introductory engineering course sequence. As part of GE 100: Fundamentals of Engineering, a course taken by every engineering freshman in his or her first semester, students develop a personal vision statement and discuss how different careers could promote the successful achievement of that vision. These personal vision statements are assigned as a formal writing assignment in the course, and they are then given to the students' advisors, who keep them on file for future reference in academic and professional advising sessions. In addition, one of the last class periods of this course is reserved for a presentation from the Dean of Engineering, who speaks to the students on "Questions for Reflecting on Your Own Sense of Vocation." In this way, the students see that no less an authority than the leader of the College of Engineering considers questions of discernment and vocation to be of such central importance that it is the only topic about which he speaks to the students in this course.

Throughout their sophomore year, students are exposed to opportunities to see how their work can fit into the broader context of a complex global society. For example, sophomore electrical and computer engineering students are asked to read a popular book such as *Soul of a New Machine* or *The World is Flat,* discuss the book in class, and then write a paper that summarizes their new understanding of how their personal calling to a technical profession impacts the lives of many other people around the world.

In their junior year, every engineering student is required to take GE 301: Principles of Professional Engineering, which is dedicated to the study of both the economic aspects of engineering work and the ethical implications of that work. As part of this course, students write four term papers, focusing on topics such as an analysis of an engineering ethics case study, the sociopolitical impact of a controversial engineering project such as the Cassini Space Probe, the impact of emerging technology on the natural environment, and an analysis of the safety characteristics of a new engineering design. These papers are designed to give students an opportunity to reflect on the way in which their work fits into a larger global context, one in which very complex interactions can lead to unexpected results. It is also hoped that the discussions in this course will lead students to fully appreciate that they have been endowed by God with not only special abilities that enable them to become an engineer, but that those abilities are accompanied by responsibilities to use them for the benefit of humankind rather than to its detriment. This course was significantly modified in 2002 as a result of the course development grant program described in the previous section.

Students' ultimate opportunity to consider engineering as a vocation comes in their senior year, when every student completes a culminating senior design project. For mechanical, electrical, and computer engineers, this is a yearlong course sequence, the Multidisciplinary Senior Design Project (GE 497 and GE 498). For civil engineering students, the project is performed in a single senior design course in their final semester, numbered CE 494. Students in each discipline are given the opportunity to take a professional-level engineering project from project definition through engineering design to final documentation. Students in GE 498 also develop a physical prototype of their design and perform a series of tests to confirm that it meets the original design constraints. Students are also asked to carefully consider the safety of their design and the impact it will have on society. This experience gives students an opportunity for the first time to fully understand the work of a professional engineer and to investigate the details of one or more subfields of their chosen engineering discipline. Throughout this experience, students are constantly encouraged to pursue opportunities through the Career Center to find a professional position or a graduate school placement that would match their personal vision statement, their particular areas of interest, and their vocation to a life of service as an engineer.

COCURRICULAR INITIATIVES

Students in Valparaiso University's College of Engineering have numerous cocurricular opportunities to engage in service learning. These opportunities range from small, generally local, service projects performed by student groups such as the American Society of Civil Engineers (ASCE) and Tau Beta Pi, or those performed by the student's fraternity and sorority, to the design and construction of large projects at the local, national, or even international level. This paper will focus on three major projects performed by students within the College of Engineering, which provided students with opportunities to provide service to their or other communities while simultaneously developing their engineering acumen. These three are: Engineers Without Borders (EWB) water supply project in Nakor, Kenya; Habitat for Humanity's ongoing projects in residential construction; and the student ASCE chapter's effort to return a portion of Valparaiso University's history, in the form of a pedestrian bridge, to campus.

Engineers Without Borders

From May 2004 through 2009, the Valparaiso University chapter of EWB was involved in a five-year water supply and irrigation project in a remote section of Northern Kenya. The goal of the project was to establish a safe drinking water supply for the villagers to replace the contaminated open-pit wells they have traditionally used, and to develop a simple, sustainable, cost-effective form of irrigation to allow the villagers to grow food in an area that receives less than three inches of rain a year. The project was designed by Valparaiso University students in the United States and constructed by those same students in Kenya, under the leadership of the first author of this paper, Carmine Polito and his colleague Michael Hagenberger.

Since the completion of the project in Kenya, the EWB chapter has initiated what will be another five-year project in Tanzania under the leadership of Dr. Hagenberger. This project involves the repair and improvement of an existing canal and irrigation system.

To date, nearly sixty Valparaiso University students and four faculty members have taken part in the field work related to these projects. Significantly more than that have taken part in the state-side planning and design aspects of the project.

With its national headquarters in Colorado, EWB is a nonprofit organization established to help people in developing areas worldwide with their engineering needs by involving, and simultaneously educating, internationally responsible engineering students. The projects typically include the design and construction of water, wastewater, water purification, sanitation, energy, and shelter systems. Projects are initiated by, and completed with, contributions from the host community, which is then trained to operate the systems without external assistance.[8]

Because EWB projects are intended to be self-sustaining, they are designed to be as mechanically simple as possible. Utilizing a simple project design facilitates the transfer of knowledge of the project construction and maintenance to the local people, who often lack even the most basic mechanical training. Additionally, the projects emphasize the use of materials that are available to the local people. This is done to simplify the maintenance of the project should replacement parts become necessary and to simplify its reproduction should a nearby community desire to replicate the project. This transfer of technology to the local people and the use of locally available materials are key elements to the EWB philosophy of sustainability.

This focus on developing simple sustainable systems, coupled with the process of transferring the knowledge to the local, often uneducated, people makes this approach perfect as a service-learning medium. By concentrating on relatively simple mechanical systems, it allows the students the opportunity to work through the design process at a manageable level, and by then actually constructing the project, they receive direct feedback on the practicality of their design. An additional advantage of the low level (by American standards) of technology involved is that it allows for participation by students, particularly freshmen and sophomores, who may not yet have obtained much engineering knowledge.

The requirement to transfer knowledge about the systems to the local people is, in itself, a fantastic learning opportunity for the students. By being forced to explain why and how the chosen system works, the students themselves develop a substantially better understanding of the materials and processes involved.

Habitat for Humanity

Another opportunity for students to engage in service is through the university's chapter of Habitat for Humanity. Nationally recognized, Habitat for Humanity constructs houses for needy members of the local community, who in turn provide assistance in the construction and then pay a mortgage on the property, which is used to fund subsequent projects.[9]

Students who participate in these construction projects, especially those who are civil engineering majors, receive valuable experience in construction processes. Founded in 1998, the university chapter of Habitat For Humanity has constructed five houses in the city of Valparaiso. Additionally, approximately twenty students, many of them engineering majors, take part in a spring break building project at a location outside of the local area. Recent projects have taken place in Mississippi, Louisiana, and Georgia.

The Student Bridge Project

The Kissing Bridge, named for a popular undergraduate activity that once took place there, spanned a series of railroad tracks on the campus's west side from the 1880s through the early 1960s. Dismantled due to age and associated safety concerns, the steel center section was sold to a local restaurateur who used it to span a creek adjacent to his establishment.[10]

Beginning in 2002, the local student chapter of the ASCE, with the help of the VU Alumni Association, purchased the steel center section of the bridge and reassembled the bridge in a new location on campus, near the entrance to the university. Part of the requirements for the project was the design and construction of a replacement bridge over the stream the bridge had recently spanned. This work, as well as the design of the foundations and abutments at the new bridge site was performed almost entirely by the students, under the guidance of Dr. Ken Leitch and Dr. Michael Hagenberger of the Civil Engineering Department.

VALPARAISO UNIVERSITY: A CULTURE OF ENGINEERING AS A VOCATION

Valparaiso University has a rich heritage of combining reason and faith, bringing together the best of both Athens and Jerusalem. The university's motto, "In Thy light we see light," illustrates the combination of faith and the scholarship that is the hallmark of a great Christian university. The College of Engineering focuses on developing students who view engineering as a vocation, who desire to work in a collaborative environment, and who are prepared to become the leaders of tomorrow.

The university puts a great deal of effort into ensuring that every student of every discipline has the opportunity to understand and apply the concepts of vocation and discernment. The Lilly Foundation has supported this effort through the Project on Theological Exploration of Vocation, which provides many opportunities for both students and faculty to consider and discuss the ideas associated with vocation. Institutional programs are in place for new and experienced faculty to learn more about their vocation as educators and scholars, course development grants are available to encourage the integration of vocation topics throughout the curricula of every department, and special workshops are arranged to help academic advisors be more than just course selection consultants.

Engineering students learn more about vocations in their Core courses, their Fundamentals of Engineering course, their Principles of Engineering Practice course, and their senior design project courses. Future curricular opportunities to promote vocations include the Campus Construction Educational Partnership and the Master of Engineering Management.

Cocurricular opportunities to pursue vocations and service learning are highlighted by Engineers Without Borders, which gives students an opportunity to apply their engineering skills internationally; Habitat for Humanity, in which they work both locally and nationally; and the Student Bridge Project, which has given students an opportunity to beautify campus while applying their engineering knowledge. Each of these opportunities helps students to see that engineering is not just an opportunity to earn a lucrative salary, but it is also an opportunity to make the world a better place through the diligent application of their God-given abilities.

NOTES

1. Valparaiso University General Catalog 2005–2006, 5–7.

2. Quoted in J. Strietelmeier, *Valparaiso's First Century* (Valparaiso, IN: Valparaiso University, 1959), 134–35.

3. M. Novak, *Business as a Calling: Work and the Examined Life* (New York: Free Press, 1996); S. R. Graves, and T. G. Addington, *Behind the Bottom Line: Powering Business Life with Spiritual Wisdom* (New York: Jossey-Bass, 2003); C. Arena, *Cause for Success* (Novato, CA: New World Library, 2004); M. Sashkin and M. G. Sashkin, *Leadership That Matters* (San Francisco: Berrett-Koehler, 2003); K. Senske, *Executive Values: A Christian Approach to Organizational Leadership* (Minneapolis: Augsburg Books, 2003).

4. See the home page for the Project on Theological Exploration of Vocation at http://www.valpo.edu/vocation/index.html.

5. Seminar on Vocation at Cambridge University, Valparaiso University Campus Wide Programs, http://www.valpo.edu/vocation/cw-seminar-cambridge.html.

6. Seminar on Vocation at Cambridge University, Valparaiso University Campus Wide Programs, http://www.valpo.edu/vocation/cw-advising-workshop.html.

7. Seminar on Vocation at Cambridge University, Valparaiso University Campus Wide Programs, http://www.valpo.edu/vocation/cw-core-faculty.html.

8. Engineers Without Borders-USA, http://www.ewb-usa.com/.

9. Valpo Habitat for Humanity, http://www.valpo.edu/student/habitat/.

10. American Society of Civil Engineers: Valpo, http://www.valpo.edu/student/asce/bridge/.

List of Contributors

JOHN BORG is an associate professor within the Department of Mechanical Engineering at Marquette University. He came to Marquette in 2002, where his primary area of research involves shock physics phenomenology. Dr. Borg has been involved in international service learning at Marquette since 2004. Formerly, Dr. Borg was a lead scientist for the Naval Surface Warfare Center in Dahlgren, Virginia, from 1997 to 2002. Dr. Borg was an NSF International Postdoctoral Fellow at Cambridge University from 1996 to 1997. He received his B.S. degree in mechanical engineering from the University of Memphis in 1990, his M.S. in aerospace engineering from the University of Notre Dame in 1992, and his Ph.D. in mechanical engineering from the University of Massachusetts Amherst in 1996.

CAMILLE M. GEORGE is an assistant professor in the School of Engineering at the University of St. Thomas in St. Paul, Minnesota. Dr. George has a broad background in applied industrial research with a particular field of expertise in the coupling of electromagnetic fields with ionized gases. Applications range from welding and surface coatings to laser cutting. She teaches an innovative class on fuel cells and is interested in energy systems and energy policy. She is also interested in global sustainability and engineering for the developing world. Dr. George has taken students to the Caribbean and is currently working with a group of educators interested in international discovery projects in the nation of Mali.

KEVIN P. HALLINAN is professor in the Department of Mechanical and Aerospace Engineering at the University of Dayton. He formerly served as chair for nearly twelve years. He received a B.S. in 1982 from the University of Akron, an M.S. in 1984 from Purdue University, and a Ph.D. in 1988 from the Johns Hopkins University, all in mechanical engineering. His teaching focus is in the areas of sustainability and energy. His research addressed these same topics, but with an emphasis on community-based energy reduction. He has authored and co-authored over eighty publications and graduated eleven Ph.D. students. He also founded a flourishing graduate program in Renewable and Clean Energy.

FR. JAMES L. HEFT, S.M., received his doctorate in historical theology from the University of Toronto. For nearly thirty years he taught at the University of Dayton where he served as chair of the Religious Studies Department, Provost, and University Professor and Chancellor. In 2006, he became the Alton Brooks Professor of Religion and President of the Institute for Advanced Catholic Studies at the University of Southern California. He has written or edited twelve books and published over 170 articles and book chapters. His most recent publications are *Catholic High Schools: Facing the New Realities* (Oxford, 2011), and an edited work, *Learned Ignorance: Intellectual Humility Among Jews, Christians and Muslims* (Oxford, 2011). Recipient of four honorary degrees, he received the 2011 Theodore M. Hesburgh Award from the Association of Catholic Colleges and Universities.

PAUL C. HEIDEBRECHT is the director of the Mennonite Central Committee (MCC) advocacy office in Ottawa, Ontario, Canada. Heidebrecht's main research interests include political theology and applying theological ethics to problems posed by technology. He earned a B.A.Sc. in mechanical engineering from the University of Waterloo in Ontario, an M.A.T.S. in theological ethics from the Associated Mennonite Biblical Seminary in Elkhart, Indiana, and a Ph.D. in religious studies from Marquette University. Prior to pursuing graduate studies he spent six years working as an engineer in the automotive industry. He has also served overseas with MCC as an appropriate technology engineer in Bangladesh and as a theology lecturer in Nigeria.

BRAD J. KALLENBERG holds a Ph.D. in theology (philosophical theology and ethics) from Fuller Theological Seminary and is associate professor of religious studies at the University of Dayton. He has taught hundreds of engineering students a course entitled Christian Ethics and Engineering. In 2005 he won the prestigious Humanities Fellowship Award to develop a course that teaches engineering design from the vantage of Christian theology and ethics. He is author of a number of scholarly articles and book chapters. He is also the author of *God and Gadgets* (Cascade Books, 2011), *Ethics as Grammar: Changing the Postmodern Subject* (University of Notre Dame Press, 2001) and is currently finishing *Living by Design: A Design Paradigm for Theological Ethics and Engineering* (Cascade Books, forthcoming).

DAVID J. O'BRIEN is professor of history and Loyola Professor of Roman Catholic Studies at the College of the Holy Cross. He has written extensively on US Catholic history, on Catholic social and political thought, and on Catholic higher education. His major publications include: *American Catholics and Social Reform: The New Deal Years* (Oxford, 1968), *From the Heart of the American Catholic Church* (Orbis, 1994), and *Public Catholicism* (Macmillan, 1988). O'Brien has served as president of the American Catholic Historical Association and has been awarded six honorary degrees. In 1992 he received the Theodore M. Hesburgh Award for Distinguished Service to Catholic Higher Education from the Association of Catholic Colleges and Universities. In 1976 he served on the staff of the National Conference of Catholic Bishops preparing for the landmark Call to Action Conference.

KRAIG J. OLEJNICZAK became dean of the College of Engineering at Valparaiso University in August 2002. He earned his B.S. in electrical engineering from Valparaiso University in 1987, and his M.S. and Ph.D. from Purdue University in 1988 and 1991, respectively. In 1991, he joined the Department of Electrical Engineering at the University of Arkansas. He led the university's High Density Electronics Center's effort in the power electronic miniaturization and packaging thrust area. Dr. Olejniczak was the recipient of the Arkansas Academy of Electrical Engineering Outstanding Faculty Award (1995 and 1999), the College of Engineering's Phillips Petroleum Outstanding Faculty Award (1996), and the IEEE Power Engineering Society's Walter Fee Award for the outstanding young power engineer in the society under the age of thirty-five.

MARGARET F. PINNELL is an associate professor in the Department of Mechanical and Aerospace Engineering and assistant dean for the School of Engineering at the University of Dayton. She teaches undergraduate and graduate materials courses and labs. Her research interests include service learning and pedagogy and materials testing and analysis. Prior to joining the School of Engineering, Dr. Pinnell worked at the University of Dayton Research Institute in the Structural Test Laboratory and at the Composites Branch of the Materials Laboratory at Wright Patterson Air Force Base.

DANIEL A. PITT is the former dean of the School of Engineering at Santa Clara University. He spent twenty-three years in technical, managerial, and executive roles at IBM (Raleigh and Zurich), Hewlett-Packard Labs (Palo Alto), Bay Networks (Santa Clara), and Nortel Networks (Santa Clara) before joining Santa Clara University. He taught as an adjunct professor for ten years at Duke University and the University of North Carolina and has served on advisory boards at the University of California, Berkeley, and the Swiss Federal Institute of Technology, Lausanne.

CARMINE POLITO is the Frederick F. Jenny Jr. Professor of Emerging Technology and an assistant professor of Civil Engineering at Valparaiso University. Professor Polito serves as the faculty adviser to the university's chapter of Engineers Without Borders. Since March of 2004 he has supervised twenty-five students on three separate trips to Kenya. He holds degrees from California Polytechnic State University, San Luis Obispo, and the Virginia Polytechnic Institute and State University. He has taught at Valparaiso University since 2001. Previously he taught at Clarkson University in Potsdam, New York, for one and a half years. He is a registered professional engineer in Indiana and California, and has worked for both CH2M HILL and the U.S. Army Corps of Engineers.

HAMID A. RAFIZADEH is professor of business at Bluffton University in Ohio. Dr. Rafizadeh teaches management decision making and efficient use of finance and economics tools and methodologies. His current research interest is the pursuit of peace and nonviolence through analysis of actions of executive management. From this perspective he seeks clarification of the link between wisdom of the sacred texts and business world activities. He has analyzed, managed, and led many industrial projects with companies in different countries including Japan, England, India, Czechoslovakia, Australia, Germany, Pakistan, Egypt, Denmark, and France. From this extensive base of experience Dr. Rafizadeh extracts the real situations he analyzes from ethical management and peacemaking perspectives. Further, he has taught and researched at seven US and international universities and brings to teaching and research an extensive base of management experience.

BARBARA K. SAIN teaches systematic theology at the University of St. Thomas in St. Paul, Minnesota. She received a Ph.D. in theology from the Catholic University of America with a dissertation on the theme of truth in the work of the twentieth-century Catholic theologian Hans Urs von Balthasar. In addition to interdisciplinary work with engineers, her research interests include theological anthropology and topics related to the communication of knowledge.

JAME SCHAEFER, associate professor of systematic theology and ethics at Marquette University, focuses her research and teaching on relating theology, the natural sciences, and technology with special attention to religious foundations for ecological ethics. Her publications include *Theological Foundations for Environmental Ethics* (Georgetown University Press, 2009) and *Confronting the Climate Crisis* (Marquette University Press, 2011). She worked with faculty of other pertinent disciplines to develop the Interdisciplinary Minor in Environmental Ethics for which she serves as director, co-steers a multidisciplinary faculty discussion group on issues that interface theology, science, and technology, and advises Students for an Environmentally Active Campus. Among her current research projects are constructing principles for a water ethic grounded in the teachings and traditions of the major world religions; formulating theologically based ethical guidance for biotechnology research, development, and commercialization; and conceptualizing a cogent theological anthropology for our age of rapidly advancing science and technology.

SCOTT J. SCHNEIDER is an assistant professor of electrical and computer engineering technology at the University of Dayton. Mr. Schneider teaches within the areas of digital electronics, software programming, computer architecture, and embedded systems. He is also interested in the vocational aspects of the engineering profession. He is currently performing research focused on developing students' awareness of engineering as a vocation. Mr. Schneider's industrial and research experience includes the development of embedded systems for both data communications and automotive control applications.

JOHN M. STAUDENMAIER, S.J., has lived and worked for thirty-one years at the University of Detroit Mercy: from 1981 to 2001 in the history department (history of technology and engineering ethics); from 1995 to 2010 as editor-in-chief of *Technology and Culture*; from 2001 to 2004 as interim dean of the College of Liberal Arts and Education; since 2005 as Assistant to the President for Mission & Identity; and from 2010 as a trustee of the university. He served as visiting professor at MIT's Science, Technology and Society Program (1982, '83, '88, '90); as research fellow at MIT's Dibner Institute (1993); as Gasson Professor at Boston College (1998–2000); and as visiting scholar at Santa Clara University's Center for Science Technology and Society (2004–5). He speaks frequently in the US and overseas, sometimes in the academy and sometimes in faith-based contexts. He also consults with museums about exhibits, with television producers about historical programs, and with science and technology reporters about articles in process. A short sample of published works suggests the kinds of questions that attract his attention: *Technology's Storytellers: Reweaving the Human Fabric* (MIT Press, 1985); "The Politics and Ethics of Engineering"; "Denying the Holy Dark: The Enlightenment and the European Mystical Tradition"; and "Rationality vs. Contingency in the History of Technology."

DOUGLAS TOUGAW is the Leitha and Willard Richardson Professor of Engineering and a professor of Electrical and Computer Engineering at Valparaiso University. Professor Tougaw's research focuses on nanotechnology and the development of advanced computer architectures composed of quantum mechanical devices. He has degrees from Rose-Hulman Institute of Technology and the University of Notre Dame. Professor Tougaw was named Outstanding Researcher of the Year by the Northwest Indiana section of Sigma Xi in 1999, and he received Honorable Mention as the Eta Kappa Nu Outstanding Young Electrical Engineer of the Year for 2005. Professor Tougaw's vocation is teaching undergraduate students. During his career, he has taught twenty-nine different courses at all levels from the freshman to the senior year and to students of all engineering disciplines. In 2005, he was named Outstanding Engineering Educator of the year by the Illinois/Indiana section of the American Society of Engineering Education (ASEE), and he currently serves on the national board of directors for ASEE.

DANIEL H. ZITOMER is professor of civil and environmental engineering and director of the Water Quality Center at Marquette University. Dr. Zitomer performs research on wastewater and drinking water treatment. He is also interested in the overlap of engineering practice and culture, focusing on linkages between international development, engineering, the humanities, ethics, and education. He led the water team for the US Agency for International Development (USAID) project Dietetics and Small Garden Systems (SGS) to Support Antiretroviral Treatment for Families Impacted by HIV/AIDS in Kenya, and other projects related to international service learning in civil and environmental engineering. He received his undergraduate degree in civil engineering from Drexel University (1988), and M.S. and Ph.D. degrees in environmental and water resources engineering from Vanderbilt University (1991, 1994).

Index